Readings from
American Scientist

Climates Past and Present

Edited by

Brian J. Skinner

Yale University

Member, Board of Editors

American Scientist

WILLIAM KAUFMANN, INC.   LOS ALTOS, CALIFORNIA

Front cover: Cardón cactus on San Pedro Mártar Island overlooking the
Sea of Cortez, Mexico. Climatic factors, such as atmospheric circulation, cause some of the world's most arid deserts to front virtually on
the shores of its largest bodies of water. Photograph by Bruce Wilcox,
Stanford University/BPS.

Contents

Introducing

Earth and Its Inhabitants

A new series of books containing readings originally published in *American Scientist*.

The 20th century has been a period of extraordinary activity for all of the sciences. During the first third of the century the greatest advances tended to be in physics; the second third was a period during which biology, and particularly molecular biology, seized the limelight; the closing third of the century is increasingly focused on the earth sciences. A sense of challenge and a growing excitement is everywhere evident in the earth sciences—especially in the papers published in *American Scientist*. With dramatic discoveries in space and the chance to compare Earth to other rocky planets, with the excitement of plate tectonics, of drifting continents and new discoveries about the evolution of environments, with a growing human population and ever increasing pressures on resources and living space, the problems facing earth sciences are growing in complexity rather than declining. We can be sure the current surge of exciting discoveries and challenges will continue to swell.

Written as a means of communicating with colleagues and students in the scientific community at large, papers in *American Scientist* are authoritative statements by specialists. Because they are meant to be read beyond the bounds of the author's special discipline, the papers do not assume a detailed knowledge on the part of the reader, are relatively free from jargon, and are generously illustrated. The papers can be read and enjoyed by any educated person. For these reasons the editors of *American Scientist* have selected a number of especially interesting papers published in recent years and grouped them into this series of topical books for use by nonspecialists and beginning students.

Each book contains ten or more articles bearing on a general theme and, though each book stands alone, it is related to and can be read in conjunction with others in the series. Traditionally the physical world has been considered to be different and separate from the biological. Physical geology, climatology, and mineral resources seemed remote from anthropology and paleontology. But a growing world population is producing anthropogenic effects that are starting to rival nature's effects in their

magnitude, and as we study these phenomena it becomes increasingly apparent that the environment we now influence has shaped us to be what we are. There is no clear boundary between the physical and the biological realms where the Earth is concerned and so the volumes in this series range from geology and geophysics to paleontology and environmental studies; taken together, they offer an authoritative modern introduction to the earth sciences.

Volumes in this Series

The Solar System and Its Strange Objects Papers discussing the origin of chemical elements, the development of planets, comets, and other objects in space; the Earth viewed from space and the place of man in the Universe.

Earth's History, Structure and Materials Readings about Earth's evolution and the way geological time is measured; also papers on plate tectonics, drifting continents and special features such as chains of volcanoes.

Climates Past and Present The record of climatic variations as read from the geological record and the factors that control the climate today, including human influence.

Earth's Energy and Mineral Resources The varieties, magnitudes, distributions, origins and exploitation of mineral and energy resources.

Paleontology and Paleoenvironments Vertebrate and invertebrate paleontology, including papers on evolutionary changes as deduced from paleontological evidence.

Evolution of Man and His Communities Hominid paleontology, paleoanthropology, and archaeology of resources in the Old World and New.

Use and Misuse of Earth's Surface Readings about the way we change and influence the environment in which we live.

Introduction: Climates Past and Present

Climate is the average condition of the weather at a particular place over a long period of time. The climate is expressed in terms of the means, extremes, and frequencies of such familiar weather elements as temperature, humidity, precipitation, and wind velocity. Climate is not constant, and fluctuations on a number of different time scales are known. Some, such as the fluctuations that produce glacial and interglacial periods, occur at about 100,000-year intervals. Others are sufficiently short-lived that one or more fluctuations have been recorded during historical times. Clearly, therefore, climatic fluctuations have more than a single cause, such as the shape of the orbit of Earth around the sun.

All climates are a result of the way that solar energy reaches Earth and is distributed by the ocean and the atmosphere. The first three articles in this volume discuss the controlling effects of the ocean and the atmosphere. Newell's discussion of climate and the ocean is particularly interesting because it proposes a way in which certain future climatic changes can be anticipated by measurement of the heat stored in the surface-water layers of the ocean. Lorenz discusses that most important weather element, the atmosphere and its circulation. The papers by Newell and Lorenz should be read together because they deal with questions of dynamics—how and why air masses flow and transfer heat as they do. Garrels, Lerman, and Mackenzie, by contrast, deal with long-term compositional changes in the atmosphere. Carbon dioxide, in particular, plays a vital role in the retention of solar heat; this gas is relatively opaque to the long-wavelength heat radiation emitted by the earth. Garrels and coworkers do not discuss climate specifically; rather, they explain the complex system of fluxes and reservoirs that controls and balances the oxygen and carbon dioxide contents of the atmosphere. There have been changes in the reservoir sizes during the past 600-million years, and these must have produced major changes in the atmosphere. This paper is, therefore, an ideal one with which to start an investigation of anthropogenic changes in the carbon dioxide content of the atmosphere.

No climatic changes are quite so dramatic as those that produce great ice masses at times of continental glaciation—the so-called "ice ages." But even though evidence for the existence of ice ages is abundant, and the advances and retreats of ice sheets can to some extent be dated, the cause of glaciation remains open to question. Beaty points out that glaciations seem not to have been caused by abrupt and dramatic changes in the global climate but were the result of a combination of factors working in concert when large land masses drifted into high latitudes.

Geologists have long known that certain kinds of sediments and sedimentary rocks were favored by specific climates. Salt deposits, for example, are formed in hot, dry climates where evaporation occurs readily, whereas coal swamps form in warm, humid conditions. Paleoclimates have traditionally been reconstructed on the basis of the residue left in the sediments. However, climatic reconstructions have created many puzzles; for instance, does the presence of coal in Antarctica mean that polar land masses have sometimes enjoyed equatorial climates? The answer, of course, lies in continental drift. As land masses moved around, the sediments they accumulated reflected the climatic zones through which they passed. Bambach, Scotese, and Ziegler interpret the climatic histories of the various continental fragments during the Paleozoic era and then use that information to recreate the geography of the Paleozoic world. Their paper is an elegant demonstration of the interplay between geology and climatology, a relationship that can be expected to bear great fruits in the future.

The closer we come to the present day, the more familiar are the flora and fauna that leave fossil traces of their presence in sediments. Because plants are especially sensitive to climate, their remains are the best indicators of climatic changes. Wolfe and Davis each draw on paleobotanical evidence in discussing climatic changes during the Tertiary and Quaternary periods, respectively.

To what extent has climate influenced history? Much more than many historians realize, perhaps. Certainly, on a long time scale involving human evolution and dispersal around the globe, climate must have played a major role, as discussed by Livingstone. But if climate has influenced us, so now are we influencing climate. Baes, Goeller, Olson, and Rotty discuss one of the most frightening (because it is uncontrolled) "experiments" in which we are all implicated—the increase in the carbon dioxide content of the atmosphere due to the burning of fossil fuels. Their paper should be read in conjunction with the earlier perceptive paper by Garrels and coworkers. Changes in the carbon dioxide content of the atmosphere are only part of the problem, however; compounds of nitrogen and sulfur are also present, as are numerous solid particles and fine liquid droplets. Panofsky discusses the general problem of air pollution meteorology, or how the atmosphere influences the distribution of polluting materials; and Middleton presents numerical evidence for the actual levels of materials in the atmosphere in a monitored industrial area. That, of course, is only part of the problem; the other part concerns the ways in which the polluting constituents actually influence and change the climate. This question is addressed by Fennelly and by Toon and Pollack.

The prediction of future climates remains elusive, intriguing, and obviously the most important question in climatology. However, we can be reasonably sure that the answer, when it comes, will derive in part from our understanding of the factors that currently control climate and in part from the paleoclimatic record.

Suggestions for
Further Reading

Bray, J. R., "Pleistocene volcanism and glacial initiation," *Science* 197: 251–254, 1977.

Budyko, M. I., *Climatic Changes* (Translation of 1974 Russian edition; Washington, D.C.: American Geophysical Union, 1977).

Cloud, P. E., Jr., "Major features of crustal evolution," *Transaction of Geological Society of South Africa;* Annex to vol. 79, Alex. L. du Toit Memorial Lecture, 1976.

Flint, R. F., *Glacial and Quaternary Geology* (New York: John Wiley and Sons, 1971).

Frakes, L. A., *Climates Throughout Geologic Time* (Amsterdam, Holland: Elsevier, 1979).

Hargraves, R. B., "Precambrian geologic history," *Science* 193: 363–371, 1976.

Hays, J. D., J. Imbrie and N. J. Shackelton, "Variations in the Earth's Orbit: Pacemaker of the Ice Ages," *Science* 194: 1121–1132, 1976.

Holland, H. D., *The Chemistry of the Atmosphere and Oceans* (New York: John Wiley and Sons, 1978).

Imbrie, J., and Katherine Palmer Imbrie, *Ice Ages: Solving the Mystery* (Short Hills, New Jersey: Enslow Publishers, 1979).

Lamb, H. H., *Climate: Present, Past and Future* (London: Methuen; Vol. I, *Fundamentals and Climate Now,* 1972; Vol. II, *Climatic History and the Future,* 1977).

Neuberger, H., and John Cahir, *Principles of Climatology* (New York: Holt, Rinehart and Winston, 1969).

Pearson, R., *Climate and Evolution* (New York: Academic Press, 1978).

Authoritative and up-to-date reviews, summaries, and analyses of many of the topics discussed in this volume can be found in the volumes published by Annual Reviews, Inc., Palo Alto, California 94306. Articles of special interest will be found in the annual volumes of *Annual Review of Earth and Planetary Sciences,* commencing with Volume 1, 1973, and *Annual Review of Fluid Mechanics,* commencing with Volume 1, 1968.

PART 1 *Climatic Controls*

Climate and the Ocean

Measurements of changes in sea-surface temperature should permit us to forecast certain climatic changes several months ahead

Reginald E. Newell

Climate is frequently defined as average weather with the averaging period taken as a month or more. For example, climate for a year might be described by a sequence of compounded monthly values for daily mean temperature or rainfall. Averaging over a month avoids the large fluctuations in weather that occur when transient systems, such as the cyclones and anticyclones of middle latitudes, pass over a given place. So-called climatic normals are often formed by collecting thirty years' worth of data for each month and forming further averages, although this procedure is a little unreliable, as there are sometimes significant fluctuations within thirty years.

In fact, climatic fluctuations occur on a variety of time scales. There are the ice ages, when the volume of water locked up as ice triples and temperature drops 10°C or more in middle latitudes; they have occurred about every 100,000 years for the past 700,000 years and perhaps more frequently before that (Shackleton and Opdyke 1976). The ice ages are thought to be caused by variations in the earth's orbit, which cause con-

Reginald E. Newell is Professor of Meteorology at M.I.T. and president of the International Commission on Climate, of the International Association of Meteorology and Atmospheric Physics. After graduating in physics from the University of Birmingham, England, he came to the United States in 1954 and took an Sc.D. at M.I.T. For 10 years he studied general atmospheric circulation, the physics of the upper atmosphere, and global pollution problems. In 1971, on a sabbatical at Imperial College, London, he took up the study of past climates and climatic fluctuations. His present work is on ocean-atmosphere interactions and the prospects for climatic forecasting. Address: Department of Meterology, 54-1520 M.I.T., Cambridge, MA 02139.

comitant variations in the pattern of solar radiation received on earth, as originally worked out by Milankovitch. This view has been substantiated recently by measurements of a series of environmental parameters in deep-sea cores, which show time variations similar to those predicted by the Milankovitch theory (see Hays et al. 1976).

Another fluctuation, which occurred in Europe and perhaps elsewhere, was that from a warmer-than-present period in the thirteenth century to a colder-than-present period centered in the seventeenth century—the latter often being termed the Little Ice Age (Lamb 1977). The amplitude of the change in the mean air temperature was perhaps only 1°–2°C, and the cause is not known. The possibility that such changes have occurred before is open, for earlier estimates of climate are rather limited.

On an even shorter time scale there are temporary coolings of a year or two, which may range up to 1°–2°C lower in average air temperature, after large volcanic eruptions have injected massive quantities of aerosols into the higher levels of the atmosphere, where they cannot be washed out by rain. Some of these events, like that of 1815, have had catastrophic effects on agriculture. In addition, there are year-to-year fluctuations that have no obvious causes.

Clearly, if we wish to plan for the future we must try to understand the fluctuations on all time scales. At present there is no accepted general theory of climate. Furthermore, observations of some of the critical parameters, such as the energy received

from the sun at the earth's orbit, have not yet been made over appropriate time intervals. Nevertheless, the broad framework of the climate system is emerging, and when it becomes clearer it should be possible to make reasonable hypotheses about what causes the year-to-year changes as well as observations to test these hypotheses.

Of the components of the terrestrial climate system, the oceans are clearly one of the most important. Because the specific heat per unit mass of water is about 4 times that of air, the thermal capacity of the oceans is over 1,000 times that of the atmosphere. In this paper I will stress the role of the oceans in regulating temperature and transporting energy in the climate system. In particular, since the temperature of the ocean surface is the variable most closely connected to the atmosphere, I will focus on the seasonal and nonseasonal variations in the top 100 m layer of water.

Energy balance in the climate system

Solar energy is continuously supplied to the earth, as shown schematically in Figure 1. The earth, in turn, emits radiation into space. It is assumed that the energy returned is equal to that received from the sun, although this may not be true, for there may be small, semi-permanent temperature changes in the deep ocean.

At the mean distance of the earth's orbit from the sun the solar flux is 1,360 W m^{-2}; because of the earth's rapid rotation the average energy available is one-fourth of this flux. The energy actually incident on the top of the atmosphere varies with latitude and season (see Smithsonian

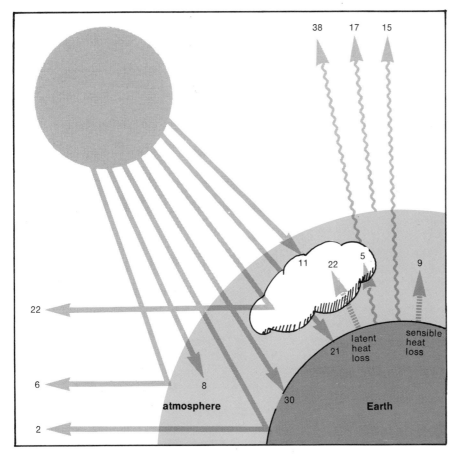

Figure 1. Solar energy incident on the earth and energy returned to space are assumed to be equal over the course of a year. Visible and infrared energy from the sun that is not reflected directly back to space from the atmosphere, the clouds, or the earth's surface is absorbed. The earth's surface loses energy mainly by evaporation from the oceans; when the water condenses and precipitates, energy is released and the clouds are thus relatively warm. Heat energy from the surface is also conducted to the air and radiated to space at infrared wavelengths (wavy lines). Infrared radiation from clouds, atmospheric gases, or the earth's surface is the mechanism by which energy that has been absorbed is lost from the earth-atmosphere system and returns to space. The total solar output is normalized to 100 units (= 340 W m^{-2}). (Values from Paltridge and Platt 1976.)

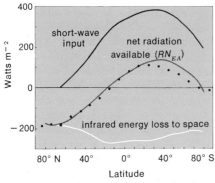

Figure 2. The difference between the short-wave input of energy from the sun and the infrared energy lost to space at the top of the atmosphere at any particular latitude is the net radiation available. The energy imbalance may be resolved either by transport of energy between latitudes and/or hemispheres or by seasonal storage and release of energy. These curves are based on climatological data for the month of January (Dopplick 1972 and pers. comm.). Black points denote satellite measurements reported by Ellis and Vonder Haar (1976).

Meteorological Tables 1963), with a maximum of about 500 W m^{-2} at high southern latitudes in summer, when the sun is above the horizon for 24 hours a day and the earth is closest to the sun.

About 30% of the incident solar energy is reflected directly back into space. The solar energy which is absorbed drives three principal energy fluxes that maintain the mean temperature of the atmosphere as a whole (Newell et al. 1974). First, the atmosphere loses heat by infrared radiation at a rate of 5.1×10^{16} W; this is energy that has been absorbed and is then radiated back at longer wavelengths. Second, the atmosphere is heated by turbulent diffusion from the land and the ocean, at about 0.7×10^{16} W. Third, the atmosphere is heated through latent heat liberation

at about 4.4×10^{16} W: energy is released when water vapor in the atmosphere condenses and precipitates. The main source of this atmospheric moisture is evaporation from the ocean—fueled, of course, by solar energy.

Water is the item of paramount importance in the atmospheric energy balance not only because latent heat is the main heat source for the air but also because infrared transfer by water vapor is the main heat sink. Clouds play a major role both in governing the amount and the path of the incoming solar radiation and in determining the infrared-energy loss to space. Most of the infrared-energy transfer depends upon the presence in the atmosphere of water vapor, carbon dioxide, and ozone, and water vapor is the most influential component. Thus, water acts both as the working substance of the atmospheric heat engine and as the cooling fluid.

Of the solar energy that reaches the surface, by far the largest fraction goes into the oceans. Their albedo—the proportion of the light received that they reflect back—is only about 10%, compared with land values that range from similar levels for certain vegetation types up to 80% for fresh snow.

If we now look at the energy balance separately for each latitude belt, we find differences between incoming solar energy and outgoing infrared radiation in each belt that indicate an energy imbalance, denoted as RN_{EA} (*n*et *r*adiation available to the *e*arth-*a*tmosphere column) in Figure 2. The summer hemisphere has more energy available than the winter hemisphere, and when annual averages are considered, low latitudes more than high.

These imbalances give rise to motions in both atmosphere and ocean that work toward smoothing them out. The motions carry energy from the summer to the winter hemisphere—making up for the deficit there—as well as from low to high latitudes. The particular types of motion involved seem to be those that most efficiently transport the energy in the required direction, though this theory of maximum efficiency has not yet been proved. In any case, it does seem that a balance is accomplished over an annual cycle, at least at low latitudes, where the average air temperature is remarkably constant from year to year.

In fact, even over the course of a single year there is little temperature change at low latitudes in response to the small variations in the incoming energy. At higher latitudes there is greater variation in temperature, for there is both a substantial variation in levels of incoming energy over the year and great variation in energy transport.

We may write an energy conservation equation for an individual latitude belt as follows:

$$RN_{EA} = S_O + S_L + S_C + S_A + S_Q + DIV(F_Q + F_A + F_O)$$

RN_{EA} represents the radiative energy available to the earth-atmosphere-ocean system in a particular latitude belt; it is the difference between the energy coming in and going out at the "top" of the atmosphere, which for practical computational purposes we may take as about 50 km. S_O represents changes in storage of heat in the whole column of the ocean; most of the change takes place in the top 300 m. S_L and S_A represent changes in storage of sensible heat in the land and in the atmosphere, respectively; S_Q represents change in storage of latent heat due to change in moisture content of the atmosphere; S_C, changes in latent-heat storage due to changes in ice volume—a variable not yet known well enough for practical use (see Oort and Vonder Haar 1976) with recent data, although it is of obvious importance on the time scale of an ice age. The divergence term (DIV) represents energy exchange with other latitude bands and has three parts: $DIV\ F_Q$, which is positive if more moisture flows out of the atmosphere of the given band than flows in, the difference being made up by an excess of evaporation over precipitation in that band; $DIV\ F_A$, which represents the sensible heat transported into or out of the band by the atmosphere; and $DIV\ F_O$, which stands for the sensible heat transported in the ocean.

The values of these various terms in the equation are given in Table 1 for several latitudes and two seasons. In the winter hemisphere, high latitudes suffer a net loss of energy to space; there is net cooling of the ocean, land,

Table 1. Components of the energy balance (in watts m^{-2})

	RN_{EA}	S_O	S_L	S_A	S_Q	$DIV(F_Q + F_A + F_O)$
Dec.–Feb.						
80°N	−146	−10	−1	−4	0	−131
60°N	−146	−29	−3	−5	0	−109
40°N	−92	−60	−2	−4	0	−27
20°N	−22	−41	−1	−1	−1	21
0°	41	0	0	0	0	41
20°S	87	49	0	2	2	36
40°S	85	74	0	3	2	8
60°S	19	30	0	3	1	−14
80°S	−36	0	2	7	0	−45
June–August						
80°N	−5	23	1	5	0	−35
60°N	32	43	3	6	1	−22
40°N	68	71	2	4	2	−11
20°N	68	30	1	1	2	35
0°	41	−10	0	0	−1	52
20°S	−24	−54	0	0	−1	32
40°S	−97	−72	0	0	0	−24
60°S	−143	−31	0	−1	0	−111
80°S	−146	0	−2	−1	0	−141

SOURCE: Newell et al. 1974.

and atmosphere; and energy is carried to high latitudes by the atmospheric and oceanic motions. At 20°N in winter, the energy lost by radiation is more than made up by energy released from the ocean, and, in fact, energy is available for export to higher latitudes. In the summer hemisphere energy is provided to all but the highest latitudes by radiative processes, much of it being stored in the sea. There is again a transport of surplus energy from low to high latitudes. Some of the energy carried to high latitudes apparently goes to increase sea temperature, as the observed increase in storage exceeds the radiative supply available at and above lat. 60°. It should be borne in mind that there are rather large uncertainties in these storage terms (±27 W m^{-2} according to Oort and Vonder Haar 1976).

When the term RN_{EA} is integrated over an entire hemisphere and over the year, there is close to an energy balance for each separate hemisphere (Vonder Haar and Suomi 1971), with no requirement for interhemispheric energy transfer. Vonder Haar and Suomi suggest that, as the Southern Hemisphere has both more ocean (whose albedo is less than that of land) and more cloud, there may be compensation between the

two effects such that the amount of energy intercepted is not detectably different from that of the Northern Hemisphere. This balance over the year does not prevail in a given season, and as has been known for some time (see, for example, Kidson et al. 1969), some of the surplus energy provided to the summer hemisphere crosses the equator to the winter hemisphere.

It is clear that the ocean, with its high thermal capacity, acts as a buffer for these energy changes. The atmosphere would cool off much more in winter if it were not for the energy taken out of the oceans. Likewise, without the ocean, atmospheric heating in summer would be very much greater. This tempering influence of the ocean is apparent when we compare the seasonal range of temperature of the air at an altitude of about 1.5 km at 50°N and 50°S: at lat. 50°N, of which 57% is covered by land, the range is 19°C, while at lat. 50°S, of which only 2% is land, the range is 6°C (Newell et al. 1972).

The difference between the responses of the land and sea surfaces to the seasonal energy change also contributes substantially to the production of north-south meanders that characterize the west-to-east atmospheric flow in the middle latitudes. In summer, the air over the land is heated much more than air over the sea, and this tends to produce areas of lower pressure over the land, of which the Asian monsoon is one example. Conversely, in winter the ocean supplies more energy to the air than does the land, and there is a tendency toward high pressure over the land. These meanderings are also influenced by mountain ranges.

The combined effects of topography and land-sea temperature differences give rise to what are often termed "standing waves" in the atmosphere above the middle latitudes. These standing waves not only contribute to the poleward energy flux themselves but also set the stage for the transient disturbances—the cyclones and anticyclones of the weather map—that are also an integral part of this flux. Again, because of the interhemispheric differences in the proportions of land and sea, the standing waves have a much smaller amplitude in the Southern Hemisphere. One of the currently intriguing problems of climatology is to try to unravel why the phase and amplitude of the standing waves vary from year to year. These variations can have profound effects on local climate, as did the temporary phase anomaly that produced a steady northerly airstream over the eastern United States in January 1977.

It is evident from the equation that the transport of energy by the ocean may be derived as a residual from the radiation, storage, and atmospheric transport terms, if these can be evaluated independently. If attention is restricted to the annual means, for which the storage terms may be set equal to zero provided no long-term trends are evident, the derivation of oceanic transports is quite simple. Radiative fluxes have been computed from climatological distributions of temperature, water vapor, carbon dioxide, ozone, and cloudiness; moreover, it has recently become possible to use satellite data to calculate the fluxes. Atmospheric transports have been derived from daily vertical sounding measurements of wind velocity, temperature, and moisture at standard pressure levels made at about 700 balloon stations in the Northern Hemisphere and at 150 in the Southern Hemisphere. An estimate of the oceanic and atmo-

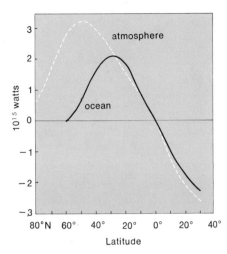

Figure 3. Both the atmosphere and the ocean transport energy poleward. Radiative imbalances are used to compute the total energy transport; then observed levels of atmospheric transport are subtracted from this total to yield the oceanic transport. Measured oceanic transport reaches maxima at about lats. 20°, while the atmospheric flux shows a maximum near lat. 40°–50° N (good values are not available for high southern latitudes). Thus, in effect, the ocean passes on energy to the atmosphere for further transport poleward. (Values from Newell et al. 1974.)

spheric fluxes appears in Figure 3. The data imply that the ocean plays an important role in the poleward transport of energy over a year.

If seasons are considered, adding the uncertainty of storage estimates, there is evidence for a transport of energy by the ocean across the equator from the summer to the winter hemisphere. Of course, the determination of the quantity of energy transported depends on knowing the quantity of heat transported by the atmosphere, and this in turn is computed from wind data that are sadly deficient for wind over the oceans. This point needs further consideration before we can refine the interhemispheric seasonal fluxes. Even without this information, however, it seems that the ocean plays an important role in the meridional transfer of energy, and it would be desirable to monitor the velocity and temperature structure of the ocean together to estimate the energy transports. At present, as we shall discuss next, only sea-surface temperatures are so monitored.

Sea-surface temperature

Sea-surface temperature has been measured and recorded on board ship for a long period; some values go back

100 years. The ships' data are clustered in certain areas, mainly freighter routes and fishing regions; they have been summarized by various groups into values for 1° or 5° grid squares, with a mean sea temperature, and usually the accompanying air temperature, calculated for each month. Using such data for the period from January 1949 to the present (except for the Indian Ocean and South Atlantic after December 1972, for which values were not available), I will outline the seasonal and nonseasonal changes in sea-surface temperature, discuss the physical factors responsible for such changes, and see if there is any information in the sea-temperature changes that might indicate future changes in mean air temperature.

Figure 4 includes maps based on the 5° × 5° grid data for February and August—the coldest and warmest months, respectively, for the oceans of the Northern Hemisphere—and a map showing the differences in ocean temperature between these months. In general, the sea is everywhere 1° or 2° warmer than the air and thus provides energy to the air. Some facets of these maps require little explanation. For example, the warmest water, that with temperature greater than about 29°C, follows the sun north and south. Yet there are large longitudinal variations in the tropics, with tongues of cold water along the equator in the eastern Atlantic and Pacific and off the Horn of Africa, that warrant a more detailed explanation.

In middle latitudes there are large east-west differences. Some of these are related to the juxtaposition of well-known ocean currents such as the cold currents from the north along the east coasts of Asia and North America and the warm Kuroshio and Gulf Stream currents from the south. These currents are thought to be driven by the wind stress, and quite good agreement has been found in some recent studies between the currents deduced from the wind-stress pattern for the North Atlantic and those actually observed (Leetma and Bunker 1978). There are also cold regions off California and northwest Africa that are not so obviously related to current patterns. And longitudinal asymmetries in the Pacific and Atlantic, well marked in the 30°–40° latitudes, occur in the open ocean well away from the coastal ef-

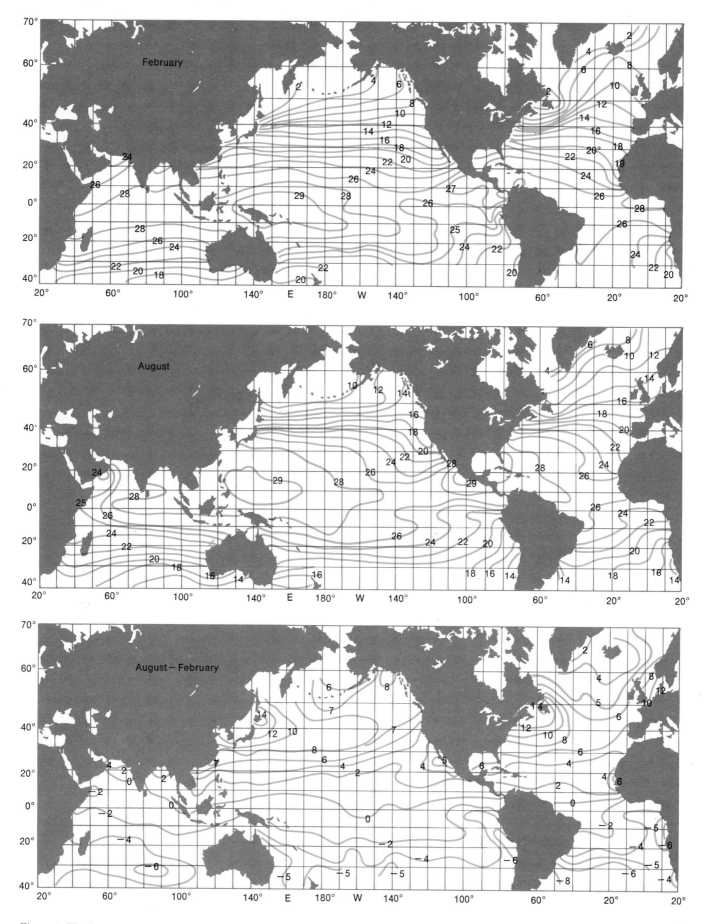

Figure 4. The long-term monthly mean sea-surface temperatures plotted (in degrees Celsius) on the maps of February and August are based on observations made from 1949 to 1972 for the Atlantic and Indian oceans and from 1963 to 1976 for the Pacific. The difference (August minus February) is shown on the map at the bottom. While water is generally warmest in the tropics, there are cool regions just to the south of the equator in the Pacific and the Atlantic and off the coasts of California, northwest Africa, Chile, and Somalia. These cool regions are thought to be caused by upwelling linked to wind-stress patterns.

Table 2. Components of the ocean-surface energy balance (in watts m^{-2})

	Evaporation	Back radiation	Sensible heat flux	Incoming solar radiation	Net
50°N, 160°W					
January	−29	−49	−10	29	−58
February	−24	−49	−6	53	−24
March	−24	−49	−3	87	15
April	−22	−44	1	121	58
May	−17	−37	10	136	92
June	−10	−27	14	121	96
July	−7	−25	12	107	87
August	−17	−29	10	103	68
September	−37	−39	4	87	15
October	−63	−53	−12	68	−53
November	−53	−56	−14	41	−78
December	−39	−49	−13	24	−73
22.5°N, 160°W					
January	−139	−63	−12	152	−62
February	−144	−63	−12	183	−36
March	−134	−60	−10	200	−3
April	−139	−56	−8	208	5
May	−118	−55	−4	236	59
June	−132	−54	−2	252	63
July	−141	−52	−1	248	52
August	−140	−53	−3	230	32
September	−138	−54	−5	219	20
October	−148	−58	−9	193	−21
November	−161	−58	−11	158	−71
December	−162	−61	−14	142	−96

SOURCE: Clark et al. 1974.

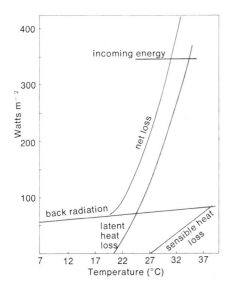

Figure 5. Tropical sea-surface temperature cannot go above about 30° C because at higher temperatures loss of energy by evaporation far exceeds energy input. The curves shown are based on a wind speed of 3 m sec^{-1} and assume that no energy is conducted downward into the ocean below the surface.

fects. Kenyon (1977) has suggested that these open-ocean asymmetries may be manifestations of the poleward transfer of energy by the oceans.

Let us consider what we know about the physical processes that maintain these observed sea-temperature patterns. For a fixed volume of ocean at a given latitude, there are three sets of processes that affect water temperature. Acting through the top are radiation, sensible heat conduction to and from the air, and loss of water and energy by evaporation. Through the sides there is advection of warm or cold water by currents. The processes that act through the base are mixing, upwelling, and downwelling.

In general, evaporation is the process through which the largest loss of energy takes place. It depends on the gradient of vapor pressure between

the surface water and the air. The evaporative energy loss may be written

$$Q_e = 6.08 (q_s - q_a) V$$

where q_s is the saturation specific humidity at the temperature of the water, q_a the specific humidity of the air, and V the wind speed. The specific humidity is defined as the ratio of the mass of water vapor to the total mass of air. With q in gm kg^{-1} and V in m sec^{-1}, Q_e is in W m^{-2}.

The sensible-heat loss likewise depends upon the temperature gradient between the sea and the air:

$$Q_s = 2.51 (T_s - T_a) V$$

where T_s is the temperature of the sea surface and T_a that of the air. Equations for both evaporation and sensible heat follow the formulations of Budyko (1974).

For the effective back radiation—the difference between long-wave radiation emitted by the sea and that coming from the atmosphere—we follow Privett (1960):

$$Q_b = 0.94 [\sigma T_s{}^4 (0.56 - 0.065 \sqrt{q_a},)]$$

where σ is the Stefan-Boltzmann constant (5.7×10^{-8} W m^{-2} K^{-4}). The saturation specific humidity q_s is mainly a function of water temperature, and if we select a value for q_a and T_a we may examine these fluxes as a function of T_s only.

To see how these equations can be used, let us take a mean value for T_a of 27°C for the tropics; this yields a q of about 15 gm kg^{-1} for a relative humidity of 70%. As a mean surface wind speed, we select 3 m sec^{-1}. Inspection of oceanic atlases (e.g. Hastenrath and Lamb 1977) shows that the mean monthly speed rarely goes below this value in the tropical regions. The three terms and their sum for a speed of 3 m sec^{-1} are plotted as a function of T_s in Figure 5. The input radiation for the tropics is about 412 W m^{-2} (Smithsonian Meteorological Tables 1963), and for an 8% oceanic albedo, no clouds, and 10% atmospheric absorption, the energy available at the surface is about 341 W m^{-2}. The energy supply and loss curves intersect at 30°–31°C where the losses are rising rather steeply with temperature, owing to the evaporation term. There is thus a natural limit on the tropical sea-surface temperature that depends basically on the radiation available and the wind speed. At temperatures 1°–2° above this limit, or at higher wind speeds, evaporative losses far exceed the possible input radiation.

In actuality, some energy is conducted down into the sea, and the solar energy is absorbed in the top few meters, not in a thin film; but even if the entire layer warms up, the temperature that can be reached is still limited by the surface loss. This limit shows up clearly in histograms of the mean monthly grid-square values of T_s (Newell et al. 1978), which have a sharp cut-off, particularly for the Indian Ocean, at about 30 °C. A similar limit on air temperature, also set by evaporation, appears over wet land—about 33 °C (Priestley 1966).

There have been a number of studies of these surface boundary terms directly; Table 2 shows some of the results of work on the North Pacific by Clark et al. (1974). In the subtropics evaporation is the principal loss, as we have already noted, and it varies comparatively little through the year. There is a net deficit of energy in winter and a net excess in the summer from solar radiation, which goes partly to increase the energy content of the water. At higher latitudes, evaporation is not the most important term. In winter, increased evaporation—due to stronger winds, which are often accompanied by stronger gradients of specific humidity—is accompanied by increased radiative losses. But again, the surface layers lose energy in winter, and there is a net gain in the summer.

Interestingly, for the latitude belt 50°–55° N in the eastern Pacific, there seems to be a net gain in energy from surface effects over the year. If this is indeed the case, there is no requirement for a flux within the ocean from

low latitudes at these longitudes. This finding contradicts the suggestion, discussed earlier, that an energy flux by the ocean is required to balance the energy budget, and obviates the need for the standing-wave mechanism discussed by Kenyon (1977). Gill and Niiler have argued before that an oceanic energy flux is not needed in the region: they found that heat input into the ocean is mainly stored locally and that horizontal advection by the mean flow is not particularly important (1973). This major problem may be resolved only with new concomitant observations of velocity and temperature.

Small-scale turbulent mixing permits temperature changes at the surface to be propagated into deeper water, as may be seen from the first two temperature profiles of Figure 6. Wind-generated turbulence often creates almost isothermal conditions in the top 20–100 m of the ocean, with a sharp temperature discontinuity,

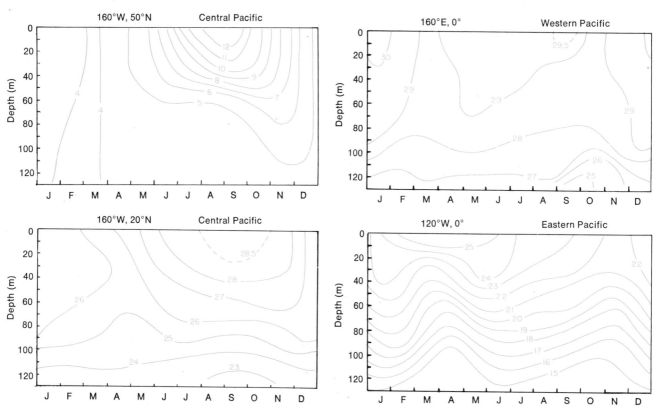

Figure 6. Temperature in the top layer of the ocean (given here in degrees Celsius) varies as a function of depth and time, as shown here for 4 places in the Pacific. The profile for the central Pacific at 50° N shows the slow downward propagation of the summer surface heating and the rapid spread of cooling with depth in winter due to mechanical mixing driven by wind stress. These effects are also present, though not so strong, at 20° N. At the equator in the western Pacific, there are 2 temperature maxima at the surface. This is to be expected, as the sun is overhead twice a year, but the timing is influenced by other factors besides solar heating in the surface energy budget. The equatorial profile in the eastern Pacific shows that the solar heating is completely overridden by upwelling in the later part of the year. The semiannual variation below 30 m has not been satisfactorily explained. (Values from Robinson and Bauer 1976.)

marking the lower limit of penetration of the turbulence. These oceanic processes are frequently interpreted in terms of one-dimensional models (see for example the papers in Kraus 1977). But the part played by quasi-horizontal processes has been explored recently by a set of observations in the Atlantic, which demonstrated the existence of almost two-dimensional eddies with characteristic lengths of about 400 km (Bretherton 1975). These will have to be included in the models and may turn out to dominate the vertical exchange processes in some regions. It will be desirable to search elsewhere for such eddies.

Another factor of major importance in controlling sea-surface temperature is upwelling caused by surface wind stress. Wind blowing across the surface of the ocean sets the water in motion, and this motion is subject to Coriolis forces—torque caused by the rotation of the earth that deflects any freely moving body to the right in the Northern Hemisphere and to the left in the Southern. In the latitude belt encompassing the equator, wind toward the west thus gives rise to poleward drift in each hemisphere, and as the top layer of water in effect moves out of the way, colder water from depth rises to take its place. Winds blowing perpendicularly toward the equator will likewise give rise to drifts in the ocean toward the west, though Coriolis forces vary in strength with latitude.

Away from the equator, the upwelling velocity may be expressed in terms of the curl of the wind stress (Stommel 1966), and maps of the velocity have been presented by Stommel (1964). At the equator itself the relationship has a discontinuity (as the Coriolis parameter reduces to zero). There is substantial upwelling, even for a wind from the east with no horizontal shear (and therefore no wind-stress curl), for there is divergence from the equatorial region in each hemisphere under the action of a westward stress and hence upwelling at the equator. Gill (1971) has derived an equation for this equatorial upwelling. The fourth profile of Figure 6 and the maps of Figure 4 show evidence of the upwelling.

These considerations of energy balance at the surface and wind-induced upwelling and turbulence provide a first-order explanation of many of the major features of the seasonal changes in sea-surface temperature fields that are evident in Figure 4. The regular variations in annual sea-surface temperature at middle and high latitudes follow variations in the solar radiation available, modified by the relatively slow conduction of energy into deeper layers in the summer and the turbulent mixing in winter that keeps the top layer isothermal. Energy that gradually enters the top layer in summer is released to the atmosphere in winter—with mixing, whether vertical or quasi-horizontal, acting to ensure that the whole top layer can lose energy through the surface.

Changes off northwestern Africa and California are brought about by stronger winds from the north in winter and concomitant drifts toward the west, which induce upwelling of colder water from below. Similar situations exist off southwestern Africa and Peru, where the winds have a southerly component. Along the equatorial Pacific and Atlantic there is also upwelling, produced by the easterly winds, and the temperature is lower directly along the equator than immediately to the north and south.

Notice that the equatorial Indian Ocean has no major cooling due to upwelling. For much of the year the equatorial winds here are from the west, south, or north rather than from the east as they are in the Pacific. Basically this is because the land-sea distribution is the main control on the surface-layer flow pattern, with movement outward from the cold continent in winter and inward toward the warm continent in summer. Upwelling produces a cold region along the coast of Somalia as the southwestern monsoon winds strengthen in early summer.

It should also be noted that the North Atlantic is warmer than the North Pacific at the same latitude and warmer than the southern oceans in the same season (see Fig. 5, and maps in Newell et al. 1978). The latter difference could be due to additional wind-enhanced turbulent mixing in the Southern Hemisphere, as the surface winds, the "roaring forties," are stronger there. The warmth of the North Atlantic water relative to that of the North Pacific is also thought to be due to differences in surface wind patterns and the concomitant curl of the wind stress, which is thought to drive the ocean currents. Leetma and Bunker (1978) show that the maximum oceanic transport in the Atlantic toward the northeast is closely related to the zero line of wind-stress curl. Because of the differences in geography and therefore surface wind, the corresponding line in the Pacific has more of an east-west orientation, and thus warm water is not carried so far north as it is in the Atlantic.

This leads us into a circular argument characteristic of air-sea studies, for the Atlantic oceanic transport has been shown to be closely related to the wind-stress pattern (Leetma et al. 1977; Leetma and Bunker 1978), while the wind in turn will depend partly on horizontal gradients of sea-surface temperature. I will not digress here into a discussion of the momentum budget of the coupled system but will just note that until this coupling is understood we cannot claim to have a complete theory.

Recent nonseasonal changes

It is obvious from the large number of factors that are involved in governing sea-surface temperature, and ultimately air temperature, that there are a myriad of possible explanations for year-to-year climatic fluctuations. Some that have been suggested are changes in the solar-energy output; changes in the phase, amplitude, and poleward energy transport by the atmospheric standing waves; changes in similar properties of oceanic standing waves; changes in the atmospheric transmission of energy, either in the visible or infrared range, due to constituent changes; and so on. One of the prime candidates has been the second theory—changes in atmospheric standing waves—and Jerome Namias, of the Scripps Institution of Oceanography, has demonstrated in a classic series of papers (1975) that sea-surface temperature anomalies, defined as departures from long-term seasonal means, are related to such atmospheric temperature- and pressure-pattern anomalies.

Namias finds that the sea-surface temperature anomalies, often of several degrees centigrade, can cover areas up to one-third or one-fourth

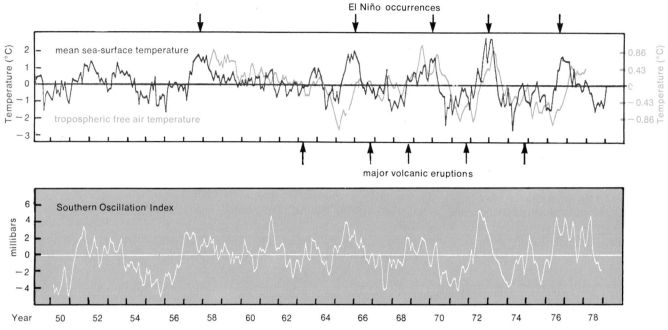

Figure 7. Changes in sea-surface temperature in the tropical Pacific precede changes in troposphere free-air temperature by a few months. The monthly mean sea temperatures are averaged over the region from long. 140° W to the coast of South America and apply to the 0°–5° S latitudinal belt. Air temperature is given as the 3-month running mean for the layer from 3–10 km altitude and is compounded from data for a series of stations near lats. 20° N and 20° S. For both air and sea, the mean seasonal cycle has been removed. Shown below is the time series of the Southern Oscillation Index taken here as the 3-month running mean of the air-pressure difference between Darwin, Australia, and Easter Island after the mean seasonal cycle has been removed; when the difference is negative, pressure at Easter Island is high and pressure at Darwin is low.

the size of the North Pacific and can persist for several months. He has studied the anomalies in relation to changing wind stress on the water and the transfer of latent and sensible heat from the water to the air and has had some success with climate forecasting (Namias 1978) based on anomaly patterns.

Clark (1967) argued that the major factor controlling the anomalies was energy-flux changes at the ocean surface, but just why these energy-flux changes should act together to produce large anomalies is a matter of some debate. Frankignoul and Hasselmann (1977), on the other hand, believe that as the transient systems of middle latitudes—the cyclones and anticyclones—pass over the ocean, they act to induce orderly sea-surface temperature anomalies through the varying wind stress on the water and the varying energy transfer between water and air. They see these small-scale, short-period atmospheric systems as yielding large-scale, long-period sea-temperature anomalies.

There is evidence to support both the view that the atmospheric changes produce oceanic anomalies and the opposite view, held, for example, by Ratcliffe and Murray (1970), who have related Atlantic sea-surface temperature anomalies to subsequent sea-level pressure anomalies.

Empirical orthogonal function analysis (Essenwanger 1976) has recently been applied to the anomalies to bring out the spatial patterns that dominate the time variability. When first applied to the North Pacific by Davis (1976), it was concluded that atmospheric changes preceded oceanic changes. Weare, Navato, and I applied this kind of analysis to a set of Pacific anomalies that extended to 20°S and found that a large fraction of the pattern variability was actually in equatorial regions, which were not included in Davis's analysis (Weare et al. 1976). Our results showed that sea-temperature changes in equatorial regions preceded tropical air-temperature changes (Newell and Weare 1976a).

To examine these equatorial changes in more detail here, we may bypass the sophisticated empirical orthogonal function analysis and use instead temperature deviations from the long-term monthly zonal means (Fig. 7). To monitor the air temperature we follow Angell and Korshover (1975) and use average temperature between pressure levels of 300 and 700 mb (corresponding to the atmospheric layer from about 3 to 10 km), for a series of stations near 20°N (3-month running mean values of this parameter also appear in Fig. 7). There are strong similarities between the curves, with the sea-temperature changes leading those of air temperature by about 3 months.

The residuals from a regression analysis of the air- and sea-temperature curves are related to the atmospheric transmission for solar energy as measured at Mauna Loa, Hawaii. The factor that is most important in governing the transmission is stratospheric aerosol from volcanic eruptions. The largest eruption during the period under consideration was that of Mt. Agung on Bali in March 1963. Our analysis of the residuals showed that this eruption reduced the tropical free-air temperature by about 0.5°C (Newell and Weare 1976b) while at the same time causing an increase in stratospheric temperature by about 5°C (Newell 1970). I think these atmospheric changes are due to absorption of solar near-infrared radiation (0.6–6 micrometer) in the stratospheric aerosol, which leads to a depletion of the near-infrared energy that normally acts to heat the troposphere. Changes in zonal mean free-air temperature at northern middle latitudes followed changes at low latitudes by several months, perhaps owing partly to the poleward diffusion of the aerosol within the stratosphere.

The major fluctuations revealed in Figure 7 are part of a global phenomenon termed the Southern Oscillation. The best-known manifestation of the oscillation is called El Niño, a tongue of warm water that sometimes appears off Peru. The warm water is particularly noticed by the Peruvian fishermen, for when it appears, as in 1965 and 1972, there is no upwelling to provide a fresh supply of nutrients, and their anchovy catch drops sharply. But basically El Niño is only one component of the Southern Oscillation.

Southern Oscillation is a term coined by Sir Gilbert Walker over fifty years ago. Walker was studying the possibility of forecasting the Indian monsoon, and, like others before him, he suggested that as the monsoon is a large-scale phenomenon, it may influence events all over the globe. He therefore made a detailed study of the correlation patterns of surface pressure, rainfall, and air and sea temperature, and found a number of significant patterns.

Walker (1932) states,

We can perhaps best sum up the situation by saying that there is a swaying of pressure on a big scale backwards and forwards between the Pacific Ocean and the Indian Ocean, and there are swayings, on a much smaller scale, between the Azores and Iceland, and between the areas of high and low pressure in the North Pacific: further, there is a marked tendency for the "highs" of the last two swayings to be accentuated when pressure in the Pacific is raised and that in the Indian Ocean lowered.

He went on to characterize these three swayings as the Southern Oscillation, the North Atlantic Oscillation, and the North Pacific Oscillation.

Walker's work on the oscillations and their application to what he termed "seasonal foreshadowing" appeared in a classic series of papers referred to as World Weather I–VI (Walker 1923–37). Walker in no way intended to imply that the oscillations were periodic. He characterized the Southern Oscillation in terms of a Southern Oscillation Index, which is taken as positive when pressure in the southeast Pacific is high, pressure in the Indian Ocean is low, and equatorial sea and air temperatures are low in both regions.

Walker incorporated pressure, temperature, and rainfall in the index, but recently pressure alone has been used, as in Figure 7. In fact, one has to take a wider view of the index, as Trenberth (1976) has stressed, because some of the correlations between station pairs change with time. Troup (1965) has drawn up a set of maps of the correlation of pressure with the Southern Oscillation Index, showing that low pressure in the tropics and subtropics from Africa to Australia and high pressure from the dateline to South America accompany a high index.

Troup suggested a toroidal circulation by which Walker's mass transfer may take place, at least between the western and eastern Pacific, with air moving from eastern to western regions in the low layers and from western to eastern in the upper troposphere. He pointed out that observed temperature differences in the equatorial Pacific were consistent with this toroidal flow. Bjerknes (1969) deduced a similar toroidal circulation and termed it "Walker circulation." In fact, toroidal circulations are not confined to the Pacific (Fig. 8).

Kidson (1975) has applied empirical orthogonal function analysis to monthly surface pressure, temperature, and precipitation departures for the tropics over the period 1951–60, and he finds spatial patterns generally in agreement with Walker's correlation analyses. He finds an additional east-west circulation cell over Brazil and the South Atlantic, similar to that proposed by Troup and Bjerknes for the Pacific, with time variations again corresponding to those of the Southern Oscillation Index.

Bjerknes showed that a warm ocean near 170°W at the equator has more rainfall than a cool ocean and suggested that the warm ocean transfers more moisture and sensible heat to the air. Referring again to Figure 7, we note that the equatorial Pacific was warmer than average in January 1973 and colder than average in January 1974. Satellite measurements showed much higher rainfall rates over the Pacific equatorial region in the former case (Rao et al. 1976), with maximum rainfall closer to the equator. But Bjerknes's suggestion is not the whole story, for Cornejo-Garrido and Stone (1977) point out that at lat. 10°S, evaporative heat loss from the ocean is higher in the eastern Pacific (under clear skies with the air dried out by the general sinking motion) than in the western Pacific, a circumstance apparently opposite to that at the equator.

Determining the cause of the large temperature and rainfall changes in the climatic system represented by the Southern Oscillation, which is next to the seasonal change in magnitude when attention is confined to the last 100 years, is obviously of paramount importance to an understanding of climatic change. Does the wind change cause the sea temperature to change? Does the sea temperature influence subsequent air temperature? Are both factors controlled by a third variable? The evidence is not yet all in, and there are many hypotheses.

The cause of El Niño, for example, has been widely debated (Barnett 1977). Perhaps the leading contender at present is the scenario developed by Wyrtki (1975). Prior to Wyrtki's analysis it was generally thought that weaker winds in the eastern Pacific—with concomitantly weaker upwelling permitting solar heating to take place—were the basic cause of El Niño. Wyrtki's contention is that there is no evidence for these weaker winds, a point still undecided. He proposes that stress produced by stronger-than-average trade winds over the central Pacific, for which he has presented data, causes a buildup of sea level in the western tropical Pacific. When the trades relax, the accumulated water flows eastward as an internal wave and leads to a thicker layer of warm water off Peru, preventing upwelling.

Wyrtki's ideas seem to fit in with the known facts about the ocean. But when air and sea temperatures are considered together, there is much evidence that fits well with earlier hypotheses. As can be seen from Figure 7, some of the extremes of the Southern Oscillation Index occur concomitantly with El Niño—indicated by the zonal mean sea temperature at 2.5°S; when the sea is warm, as in 1958, 1965, and 1972, the index as it is defined in the figure is large; this actually corresponds to a negative index as defined by Walker. There have been many suggestions that the index may be used to forecast El Niño, and a formal regression

and my colleagues are currently taking a new look at the Indian Ocean sea-temperature patterns to test the two hypotheses further. Walker found that persistence of the pressure anomalies after the summer monsoon is much stronger than after other seasons, further evidence linking the oscillation and the monsoon.

As is obvious from Figure 7, one should be able, from observations of sea temperature and volcanic aerosol,

and therefore diminished upwelling and higher ocean temperatures. At the same time, pressure is lower over northeastern Brazil, and on the average there is more rain. Thus the rain probably originates not from a locally warmer sea but from a modulation of the general tropical circulation. Nevertheless, sea-surface temperature can be used to predict the level of rainfall.

The basic reason for the Southern

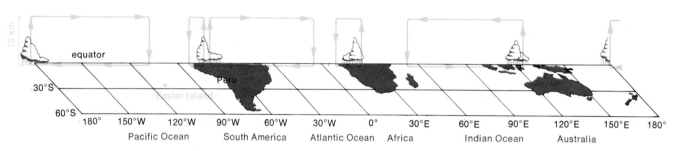

Figure 8. Vertical atmospheric circulation (called Walker circulation) is shown here as it appears over equatorial regions in January. The Southern Oscillation Index measures the strength of this circulation. If pressure is higher than normal over Darwin, then the atmospheric rising motion there is weaker than average and the sinking motion just west of Peru is weaker than average. Eastern Pacific oceanic

upwelling is weaker, too, a probable explanation being that the horizontal winds that are the lower arm of the Pacific cell are weaker, and the water off Peru may be heated up by the sun. This situation is often accompanied by heavy rain over Peru and higher than average air temperature throughout the tropics. In the opposite phase of the Southern Oscillation, when the pressure is higher than normal over

Easter Island and lower than normal over Darwin, the circulation cells in the Pacific and Indian oceans are stronger than average. There is a cold ocean off Peru owing to stronger upwelling (perhaps caused by the stronger winds) and concomitantly a cold tropical troposphere. (Data from work by Boer and Kyle; see Newell et al. 1974.)

analysis of these two curves yields maximum correlation with the sea temperature that follows the index by 2 months.

Walker was apparently right in suspecting a relationship between the Indian monsoon and the observed pressure correlations. Maps of the vertical motion accompanying the monsoon show that rising motion in the atmosphere over the eastern Indian Ocean is accompanied by sinking motion over the eastern Pacific. An enhancement of this circulation system would lead to lower pressure, more clouds, and less solar heating of the ocean over the former region, accompanied by higher pressure over the latter, or a high value of the Southern Oscillation Index on Walker's definition. Stronger winds, more upwelling, and a colder equatorial ocean could ensue in the eastern Pacific, and thus the two tropical ocean regions would be cool at the same time but for different physical reasons. These ideas are not directly compatible with Wyrtki's hypothesis,

to forecast mean temperature of the tropical troposphere 1 or 2 months ahead, even when precise causal mechanisms are not known. But it must be stressed that a few months is the real limit, because we are at present unable to forecast volcanic eruptions, and these play a major role in governing air temperature.

There are substantial variations in rainfall as well as temperature that are related to the Southern Oscillation. For example, Markham and McLain (1977) have reported that if sea-surface temperature in the Atlantic upwelling region (say 0°–5°S) is higher than average in December, rainfall in northeastern Brazil will be higher than normal in the following January–March. This correlation is, of course, only one part of the Southern Oscillation: Troup (1965) has shown that a high value of the Southern Oscillation Index is accompanied by higher-than-normal pressures in the Atlantic, which allows more radiation to be received at the surface and leads to weaker winds

Oscillation, or monsoon modulation, is not understood, but it appears that these pressure patterns delineate natural modes of the tropical atmosphere-ocean system. Theoretical and modeling studies such as the recent contributions by Webster (1973) and Rowntree (1976) will undoubtedly throw more light on the structure and physical background of these natural modes. Eventually, of course, feedback between the ocean and atmosphere will have to be included in these models, and so a modeling of the ocean itself will have to be included. Bryan and his colleagues (1975) have achieved a coupled model that describes the mean fields, but it will be necessary to understand physical processes such as two-dimensional eddies thoroughly before the coupled models can be successful.

I would argue that it is these changes in the general circulation of the atmosphere that account for most of the large changes in tropical sea and air temperatures observed since 1949, and perhaps much earlier, with volcanic eruptions accounting for a substantial fraction of the remaining variations. If this is indeed the case, there is not much hope for long-term climate forecasting until we can understand the factors governing these natural modes of the general circulation in more detail and can forecast volcanic eruptions. Once a change in the circulation is under way, however, then the various lag relationships that we are discovering should enable us to forecast successfully some simple parameters, such as the tropical mean temperature and rainfall in various regions, several months ahead.

References

Angell, J. K., and J. Korshover. 1975. *Mon. Weath. Rev.* 103:1007–12.

Barnett, T. P. 1977. *J. Phys. Oceanogr.* 7: 633–47.

Bjerknes, J. 1969. *Mon. Weath. Rev.* 97:163–72.

Bretherton, F. P. 1975. *Quart. J. Roy. Meteor. Soc.* 101:705–21.

Bryan, K., S. Manabe, and R. C. Pacanowski. 1975. *J. Phys. Oceanogr.* 5:30–46.

Budyko, M. I. 1974. *Climate and Life.* Academic Press.

Clark, N. E. 1967. Report on an Investigation of Large-Scale Heat Transfer Processes and Fluctuations of Sea-Surface Temperature in the North Pacific Ocean. Cambridge, MA: Solar Energy Research Fund, M.I.T.

Clark, N. E., L. Eber, R. M. Laurs, J. A. Renner, and J. F. T. Saur. 1974. Heat exchange between ocean and atmosphere in the eastern North Pacific for 1961–71. Seattle: N.O.A.A. Tech. Rept. NMFS SSRF-682.

Cornejo-Garrido, A. G., and P. H. Stone. 1977. *J. Atmos. Sci.* 34:1155–62.

Davis, R. E. 1976. *J. Phys. Oceanogr.* 6:249–66.

Dopplick, T. G. 1972. *J. Atmos. Sci.* 29: 1278–94.

Ellis, J. S., and T. H. Vonder Haar. 1976. Zonal Average Earth Radiation Budget Measurements from Satellites for Climate Studies. Atmos. Sci. Paper no. 240, Colorado State Univ.

Essenwanger, O. 1976. *Applied Statistics in Atmospheric Science.* Amsterdam: Elsevier.

Frankignoul, C., and K. Hasselmann. 1977. *Tellus* 29:289–305.

Gill, A. E. 1971. *Deep-Sea Res.* 18:421–31.

Gill, A. E., and P. P. Niiler. 1973. *Deep-Sea Res.* 20:141–77.

Hastenrath, S., and P. J. Lamb. 1977. *Climatic Atlas of the Tropical Atlantic and Eastern Pacific Oceans.* Univ. of Wisconsin Press.

Hays, J. D., J. Imbrie, and N. J. Shackleton. 1976. *Science* 194:1121–32.

Kenyon, K. E. 1977. *J. Phys. Oceanogr.* 7: 256–63.

Kidson, J. W. 1975. *Mon. Weath. Rev.* 103: 187–96.

Kidson, J. W., D. G. Vincent, and R. E. Newell. 1969. *Quart. J. Roy. Meteor. Soc.* 95:258–87.

Kraus, E., ed. 1977. *Modelling and Prediction of the Upper Layers of the Ocean.* Pergamon.

Lamb, H. H. 1977. *Climate: Present, Past, and Future,* vol. 2. London: Methuen.

Leetma, A., and A. F. Bunker. 1978. *J. Marine Res.* 36:311–22.

Leetma, A., P. Niiler, and H. Stommel. 1977. *J. Marine Res.* 35:1–9.

Markham, C. G., and D. R. McLain. 1977. *Nature* 265:320–23.

Namias, J. 1975. *Short-Period Climatic Variations.* Collected Works 1934–1974, vols. 1 and 2. Scripps Inst. Oceanogr.

———. 1978. *Revs. of Geophys. and Space Phys.* 16:435–58.

Newell, R. E. 1970. *Nature* 227:697–99.

Newell, R. E., J. W. Kidson, D. G. Vincent, and G. J. Boer. 1972. *The General Circulation of the Tropical Atmosphere and Interactions with Extratropical Latitudes,* vol. 1. M.I.T. Press.

———. 1974. *The General Circulation of the Tropical Atmosphere and Interactions with Extratropical Latitudes,* vol. 2. M.I.T. Press.

Newell, R. E., A. R. Navato, and J. Hsiung. 1978. *Pure and Appl. Geophys.* 116:351–71.

Newell, R. E., and B. C. Weare. 1976a. *Nature* 262:40–41.

———. 1976b. *Science* 194:1413–14.

Oort, A. H., and T. H. Vonder Haar. 1976. *J. Phys. Oceanogr.* 6:781–800.

Paltridge, G. W., and C. M. R. Platt. 1976. *Radiative Processes in Meteorology and Climatology.* Amsterdam: Elsevier.

Priestley, C. H. B. 1966. *Agr. Meteor.* 3:241–46.

Privett, D. W. 1960. *Geophys. Memoirs* 13, no. 104. London: Meteor. Office.

Rao, M. S. V., W. V. Abbott III, and J. S. Theon. 1976. Satellite-derived global oceanic rainfall atlas (1973 and 1974). Rept. NASA SP-410.

Ratcliffe, R. A. S., and R. Murray. 1970. *Quart. J. Roy. Meteor. Soc.* 96:226–46.

Robinson, M. K., and R. A. Bauer. 1976. *Atlas of North Pacific Ocean Monthly Mean Temperatures and Mean Salinities of the Surface Layer.* Washington, DC: Naval Oceanographic Office.

Rowntree, P. R. 1976. *Quart. J. Roy. Meteor. Soc.* 102:583–605.

Shackleton, N. J., and N. D. Opdyke. 1976. Investigation of Late Quaternary paleo-oceanography and paleoclimatology. In *Geol. Soc. of America Memoir 145,* ed. R. M. Cline and J. D. Hays, pp. 449–64.

Smithsonian Meteorological Tables. 1963. Sixth ed. prepared by R. J. List. Smithsonian Miscellaneous Collections, vol. 114. Smithsonian Inst.

Stommel, H. 1964. In *Studies on Oceanography,* ed. K. Yoshida, pp. 53–58. Univ. of Washington Press.

———. 1966. *The Gulf Stream.* Univ. of California Press.

Trenberth, K. E. 1976. *Quart. J. Roy. Meteor. Soc.* 102:639–53.

Troup, A. J. 1965. *Quart. J. Roy. Meteor. Soc.* 91:490–506.

Vonder Haar, T. H., and V. Suomi. 1971. *J. Atmos. Sci.* 28:305–14.

Walker, G. T. 1923. *India. Meteor. Dept. Mem.* 24:75–131.

———. 1924. *India. Meteor. Dept. Mem.* 24: 275–332.

———. 1928. *Mem. Roy. Meteor. Soc.* 2(17): 97–134.

———. 1930. *Mem. Roy. Meteor. Soc.* 3(24): 81–95.

———. 1932. *Mem. Roy. Meteor. Soc.* 4(36): 53–84.

———. 1937. *Mem. Roy. Meteor. Soc.* 6(39): 119–39.

Weare, B. C., A. R. Navato, and R. E. Newell. 1976. *J. Phys. Oceanogr.* 6:671–78.

Webster, P. J. 1973. *Mon. Weath. Rev.* 101: 803–16.

Wyrtki, K. 1975. *J. Phys. Oceanogr.* 5:572–84.

"It's not the humidity—it's the thermal pollution."

Edward N. Lorenz

The Circulation of the Atmosphere

In a high school classroom where I sat some years ago, our physics teacher once put before us the proposition that the sun is the source of all our energy. Naturally we tried to upset his claim, and I finally asked him about the ocean tides. At this point he invoked the hypothesis, which was reasonably well accepted in that day, that the planets and their satellites had at one time been part of the sun. If the sun was the source of the moon itself, it was certainly the source of the energy in the moon.

The thesis which our teacher supported to our partial satisfaction a generation ago would be much harder to defend today. For one thing, alternative hypotheses which attribute the origin of the planets to accretion rather than fragmentation have become more widely accepted. Closer to home, such devices as nuclear reactors which were then unknown are now commonplace. Yet, despite all these new developments, there remain many scientific fields where the sun may still be treated as the source of nearly all the relevant energy, and these fields continue to possess many unsolved problems which are as challenging as those encountered in some of the younger sciences. Among these fields is the study of the circulation of the earth's atmosphere.

If a fluid is subjected to non-uniform heating, a circulation will ordinarily develop. One of the problems of greatest concern to the fluid dynamicist is that of deducing from the basic law of physics the circulation which will take place in a particular fluid system when it is heated in a particular fashion. Even for some of the simplest systems—for example, a tank of water insulated on the bottom and sides and cooled at the top—the problem is but partially solved.

Trained in mathematics at Dartmouth College and Harvard University, Dr. Lorenz took an ScD degree in meteorology in 1948. As an officer in the Army Air Corps, he saw service as a weather forecaster on the islands of Saipan, Guam, and Okinawa from 1942 to 1946. As Professor of Meteorology, since 1948, at M.I. T., he has also been visiting scientist at the University of California, Los Angeles, and the Norske Meteorologiske Institutt in Oslo, Norway. This article is concerned with the motion of the atmosphere, generally organized into broad easterly and westerly currents, some of which encircle the globe. Large wave-like disturbances, often several thousand miles across, are embedded in these currents. The implications of such circulations for the future of weather forecasting are discussed. Address: Department of Meteorology, Massachusetts Institute of Technology, Cambridge, Mass. 02139.

The atmosphere itself is a special fluid, and it is heated more strongly by the sun in equatorial and tropical latitudes than in temperate and especially polar latitudes. It must therefore possess a circulation. One of the dreams of the theoretical meteorologist has been the deduction of this circulation from basic principles, given such quantities as the mass, radius, and angular velocity of the earth, the total mass and composition of the atmosphere, and the intensity and spectral distribution of the radiation from the sun.

Actually, there is some question as to whether the meteorologist can really deduce the circulation. He could undoubtedly do so if he could determine the general solution of the mathematical equations which represent the physical laws. But the exact equations are too complicated to be handled by any known method, and some simplifying approximations are essential. The meteorologist is already familiar with the general features of the circulation, and this knowledge will in all probability influence him in choosing among the many available approximations. For example, in many theoretical studies the complete three-dimensional distribution of the atmospheric variables—pressure, temperature, wind, moisture—is represented by the two-dimensional distributions of these variables at a few chosen levels, the values occurring between these levels being obtained by interpolation. But the number and spacing of the levels is generally based upon previous knowledge of the true atmospheric structure. In view of such circumstances, many meteorologists regard their task as that of explaining or accounting for the circulation as it is observed, rather than deducing it from basic considerations.

Before we consider how one might account for the observed circulation, let us attempt to put ourselves on a more nearly equal footing with the theoretical meteorologist by looking at some of its principal features. One of the first things to be noticed is that the total circulation is composed of identifiable circulation systems of widely differing horizontal scales. We shall first examine some of the smaller-scale systems, and then progress toward the larger ones.

Perhaps the smallest-scale system to be dignified by a special name is the dust devil. These whirlring columns of dust-laden air typically reach heights of a few hundred feet, but over hot deserts they sometimes extend half a mile upward. The dust serves only to make them visible;

vortices of this size often form where no dust is available, and presumably many such vortices are invisible for each one which can be seen. If a person should be struck by one, he would probably experience nothing more than a sudden gust of wind. In fact, when an unexpected gust is encountered, it is frequently a portion of a dustless dust devil. The opposing motion a few yards away may remain undetected.

Despite their general inconsequential effect upon human activity, these small systems are so numerous that collectively they comprise a significant element in the total circulation of the atmosphere. Their most obvious motion is rotary, but they also contain powerful upward currents. The dust which they raise will fall back to the ground when the circulation wanes, but in the meantime they are an effective mechanism for conveying heat from the ground to higher levels.

Occurring in many sizes, but ordinarily larger than dust devils, are cumulus clouds. Within and directly underneath the clouds the currents are mainly upward, and the compensating downward currents occur largely in and directly below the drier spaces between the clouds. Particularly in tropical latitudes, cumulus cloud circulations are one of the principal mechanisms for conveying water from the earth's surface up to higher levels in the atmosphere.

The most fully developed cumiliform clouds, the cumulonimbus, generally contain showers, and often thunderstorms. In extreme cases they are ten miles deep. In that event hail is likely, and tornado funnels may reach from the main cloud mass down to the ground.

The tornado bears a superficial resemblance to the dust devil, but the energy for maintaining it appears to come from the cloud above rather than the ground below. It is the most violent of storms, and has received much study because of its devastating effect upon human life and property. Nevertheless, tornadoes are not a very important element in the total circulation simply because there are so few of them; if a gram of air could be picked at random from the atmosphere, the probability that it would be taken from a tornado is about one in 10^{12}.

Individual cumulus clouds are seldom randomly distributed, but tend to be organized into systems of larger scale. Figure 1 is a photograph of a radar scope, showing a long line of thunderstorms extending across eastern Oklahoma. The display is arranged in the form of a map, the location of the radar being represented by the center of the large bright spot. The line extending northward from this spot is a reference line. The storms appear to the southeast.

Ordinary clouds, whose droplets are typically a hundredth of a millimeter or so in diameter, are transparent to radar rays. Only those clouds containing raindrops (or snowflakes, or hailstones), whose diameters often exceed a millimeter, show up on the scope. This particular scope is designed so that different intensities of the reflected radar signal, resulting from different intensities of rain, show up as different shadings. The brightest spots within the line indicate the heaviest rain—in this case, thunderstorms. About twenty of these storms are organized into a line about two hundred miles long.

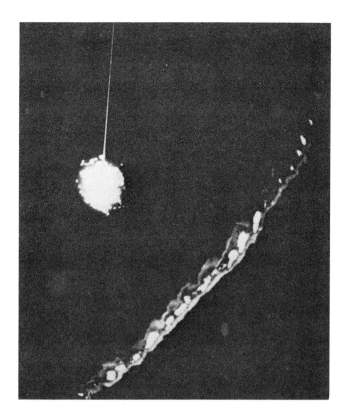

Figure 1. A line of thunderstorms at 1930 C.S.T., May 10, 1964, as seen by the WSR-57 radar of the National Severe Storms Laboratory, U. S. Weather Bureau at Norman, Oklahoma. Photo by Charles Clark, reproduced through courtesy of the *Monthly Weather Review*.

Lines of thunderstorms or heavy rain frequently form portions of still larger systems. Figure 2 is another radar photograph, which shows a tropical hurricane as it strikes southern Florida. The bright areas are again rain, but this radar does not differentiate between intensities. The dry central eye is plainly visible. Surrounding it is a very wet eye wall composed of towering cumulonimbus clouds, and a complex of spiral rain bands whose individual structures are somewhat like that of the line shown in Figure 1.

Photographing complete storms in visible light has recently been made possible by the satellite. Figure 3 shows a storm over the north Atlantic. The area covered by the photograph is nearly a thousand miles square. The storm, which is not of tropical origin, looks very much like the tropical storm of Figure 2. It should be noted, however, that in Figure 3 we are seeing clouds rather than rain. We may assume that rain is falling from some of the heavier clouds, but altogether the storm contains far less water than its tropical counterpart.

To display a larger system in a single picture we may combine information obtained from different geographical locations. Some interesting results have been obtained by piecing together successive photographs from a single satellite, but most of our composite descriptions of the atmosphere are in the form of weather maps.

Figure 4 is a northern hemisphere map for a particular late winter day. The lines are isobars—lines of constant pressure, after the pressure has been reduced to sea level by a standard procedure. The relation between the pressure field and the field of motion is given to a first

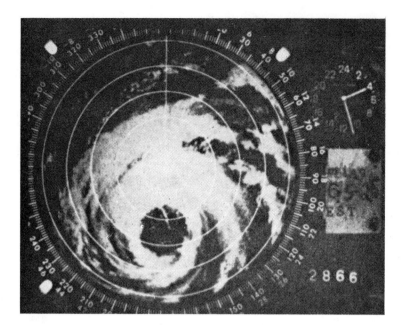

Figure 2. Hurricane Betsy at 0427 E.S.T., September 8, 1965, as seen by the WSR-57 radar of the United States Weather Bureau at Miami, Florida. U. S. Weather Bureau photo, reproduced through courtesy of the American Meteorological Society.

approximation by the geostrophic wind law. This law, which is often lesson number one to the meteorology student, states that the air moves parallel to the isobars, traveling clockwise about a high pressure area and counterclockwise about a low pressure area in the northern hemisphere, and in the opposite sense in the southern hemisphere. The direction of the geostrophic wind at sea level is indicated in Figure 4 by arrows attached to the isobars.

Those who have had experience with nonrotating fluids may be more familiar with motion at right angles to the isobars, toward lower pressure. Such motion will occur, for example, when the pressure force, which is directed

Figure 3. A north Atlantic storm, centered at 48°N, 22°W, at 1402 G.M.T., June 11, 1964, as seen by the TIROS VII weather satellite. NASA photo, reprinted through courtesy of Aracon Geophysics Co.

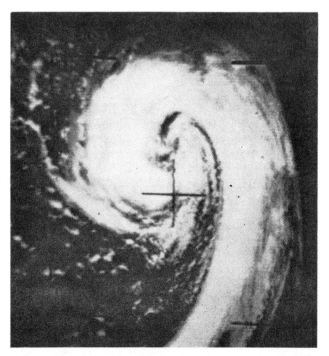

toward lower pressure, is balanced by the force of friction, which opposes the motion. Throughout most of the atmosphere, however, friction is a minor force, and the pressure force is ordinarily balanced by the Coriolis force—the deflecting force resulting from the earth's rotation—which acts at right angles to the motion. Near the ground, friction assumes a more important role, and the wind tends to have a small component toward lower pressure superposed upon the geostrophic component. A mathematical demonstration that the forces must be nearly in balance would be extremely involved; it is simply a matter of observation that the forces do tend to balance for the most part.

Referring to Figure 4 we observe that in lower latitudes a belt of easterly winds nearly encircles the globe. These easterlies are the familiar trade winds—the steadiest of the global currents. In connection with the trades we often hear of the prevailing westerlies in middle latitudes, but in Figure 4 we see no globe-encircling belt of westerlies. If we measure the wind at a sufficient number of points in middle latitudes, and then average these measurements, we shall find a resultant wind from the west, so that the prevailing westerlies are indeed present, but only in a statistical sense. They are not present at all locations at one time, nor are they present at all times at one location. The more obvious feature in middle latitudes is the great number of anticyclones and cyclones—the centers of high and low pressure with their accompanying clockwise and counterclockwise vortices. We have seen examples of cyclones in the radar and satellite photographs.

Figure 5 is similar to Figure 4 except that it presents the conditions at an elevation of about 30,000 ft. The low-latitude easterlies are less prominent than at sea level, but the middle-latitude westerlies are much more pronounced, and form a belt which encircles the globe. Cyclones and anticyclones are still present, but to some extent they have been replaced by troughs and ridges—lines along which the pressure is lower or higher than at adjacent longitudes. The strongest winds, indicated in Figure 5 by the most closely packed isobars, form a relatively narrow continuous current which nearly encircles the polar

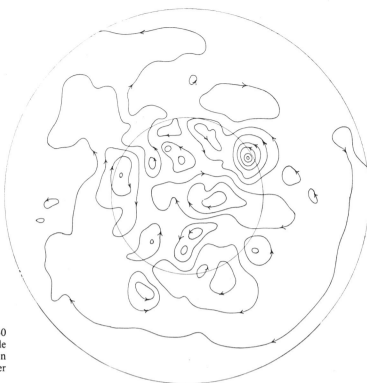

Figure 4. Northern hemisphere weather map at sea level, 1230 G.M.T., March 15, 1952. Outer circle is equator, and inner circle is 45th parallel. Heavy lines are isobars, and arrows show direction of geostrophic wind. Isobar positions based upon U. S. Weather Bureau analysis.

regions; this is the now familiar jet stream. Winds as high as 200 miles per hour are not uncommon there.

The maps in Figures 4 and 5 present conditions on a single day, but the flow which they illustrate is for the most part typical of the general behavior of the atmosphere. Important features which are not present or not clearly revealed are tropical hurricanes, which are confined mainly to the summer or autumn hemisphere, and the intertropical convergence zone, a rather narrow region extending virtually around the globe where currents from the northern and southern hemispheres converge and rise, and in doing so bring about rather heavy cumulus convection, with frequent thunderstorms.

The motions of global scale—the trade winds and the prevailing westerlies at low levels, the upper-level westerlies which culminate in the jet stream, and the intertropical convergence zone—form what is ordinarily called the general circulation of the atmosphere. The

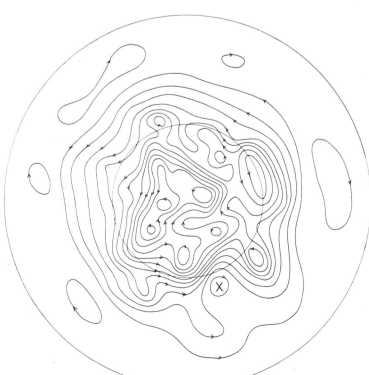

Figure 5. Northern hemisphere weather map at 30,000 feet elevation, 1000 G.M.T., March 15, 1952. Lines have same meanings as in Figure 4. Isobar positions based upon U. S. Weather Bureau analysis.

migratory cyclones and anticyclones and the accompanying upper-level troughs and ridges, and at lower latitudes the tropical hurricanes, are usually classified as secondary circulations. There is not complete agreement, however, as to how the general circulation should be defined; I personally prefer to regard at least the existence of cyclones and anticyclones, and some of their collective statistical properties, as basic characteristics of the general circulation.

Having noted some of the principal features of the circulation of the atmosphere, let us see to what extent they can be accounted for. We shall first examine some features of global scale. Proposed explanations for the trade winds date back several centuries, but Hadley (1735) was the first to recognize the significance of the earth's rotation, and his ideas have become a familiar part of meteorological history.

Hadley assumed that the excess solar heating in lower latitudes over that in higher latitudes would bring about a general rising motion in lower latitudes and sinking in higher latitudes, the circuit being completed by a general equatorward motion near the earth's surface and a poleward motion aloft. He argued that there would be no reason for systematic eastward or westward motion if the earth were not rotating. He noted, however, that the surface of a rotating earth moves more rapidly toward the east in low latitudes than in high latitudes, in an absolute sense. Air moving directly equatorward across some middle latitude would therefore, in trying to conserve its absolute eastward velocity, arrive at lower latitudes with a westward motion relative to the earth—thus the trade winds.

Hadley's numerical calculations indicated that the air would acquire much higher westward speeds than those actually observed, and he attributed the lower speeds to the frictional drag between the air and the earth's surface. He then noted that this drag would continually slow down the rotation of the earth, unless counteracted by an equal and opposite drag somewhere else; this he assumed to occur in the latitudes of the prevailing westerlies. To account for the westerlies he argued that the air which had reached low latitudes and risen to higher elevations would, upon returning poleward, acquire an eastward motion relative to the earth, thus becoming the upper-level westerlies. Upon sinking in high latitudes it would form the prevailing westerlies, after which it would again move equatorward, completing the circuit.

It is easy enough in the light of today's knowledge to find fault with Hadley's reasoning. For one thing, the equatorward or poleward moving air tends to conserve its absolute angular momentum rather than its absolute velocity, and Hadley's assumption to the contrary caused his numerical calculations to be too small by a factor of two. The tendency of the air to conserve its angular momentum is identical with what we now call the eastward component of the Coriolis force. But Hadley preceded Coriolis by a century, and it is perhaps to his credit that he was as nearly correct as he was. Moreover, his quantitative error does not invalidate his subsequent reasoning, which is entirely qualitative; his assumption concerning friction was simply that it would reduce the strength of the trade winds to what it is observed to be from what it would be otherwise.

A more general criticism of Hadley's work is that it fails to demonstrate that the circulation must assume its observed form in preference to some other, since it lacks the necessary quantitative treatment. It may be shown that in a thermally forced system the warmer air must rise and the colder air must sink in some over-all sense, or, more precisely, that the temperature and the upward motion must be positively correlated, but the correlation need not be perfect nor even very high. Hadley assumed, without justification, that all of the rising air was warmer than all of the sinking air, and his subsequent conclusions were, therefore, not demanded by the physical laws.

Nevertheless, as an essentially correct account of what does take place, as opposed to what must take place, Hadley's work went virtually unchallenged for nearly two centuries. Once the dynamical effect of the earth's rotation had been properly expressed in mathematical form (e.g., by Coriolis, 1835) it was possible to replace the qualitative reasoning by quantitative computations, and ultimately a number of theoreticians attempted to do so. As we have already noted, the exact equations proved to be so complex that numerous simplifications were required. Thus, as in Hadley's paper, the presence of oceans and continents was generally neglected. Within the framework of such simplifications, the early quantitative results (e.g., Oberbeck, 1888) generally confirmed Hadley's work.

We may reword Hadley's arguments concerning the frictional drag in terms of absolute angular momentum, by saying that angular momentum (regarded as positive if the motion is eastward) is transferred from the earth to the atmosphere in the latitudes of the trade winds, and from the atmosphere to the earth in the latitudes of the prevailing westerlies. Since the earth is to a large extent a solid, it does not acquire any differential rotation thereby, but the atmosphere, being a fluid, would continue to speed up (in the absolute sense) in low latitudes and slow down in high latitudes, were it not for some mechanism for conveying angular momentum from low to high latitudes within the atmosphere. Hadley's picture contains such a mechanism since the poleward moving air aloft carries with it more eastward angular momentum than the equatorward moving air below. Nevertheless, other mechanisms are physically possible.

A feature of Hadley's paper which was characteristic of much of the ensuing work is that the large-scale currents were assumed to behave quite independently of any secondary circulations. As one familiar with the high seas, Hadley was well aware of the violent storms which were often encountered, but he presumably looked upon them as irrelevant as far as the maintenance of the trade winds was concerned. Later investigators recognized the potential importance of the secondary circulations, but tended to regard them as a sort of large-scale turbulence, which could be suitably incorporated into the mathematical equations by choosing larger coefficients of viscosity and thermal conductivity than would otherwise be demanded. Since the appropriate values of these coefficients were not known in any case, the inclusion of storms would not invalidate any results which had been previously arrived at.

But sooner or later all theories are challenged, and Jeffreys (1926) eventually proposed that the secondary circulations were actually responsible for maintaining the global currents. This idea was not well received by those who based their reasoning upon turbulence theory; turbulence should tend to smooth out the temperature field by conveying heat from latitudes of high to those of low temperature; likewise, it should tend to create a state of solid rotation by conveying angular momentum from latitudes of high to those of low angular velocity. Jeffreys was proposing that, unlike turbulence, the secondary circulations conveyed angular momentum in the opposite direction.

For a number of years Jeffreys' ideas were no more than alternatives to Hadley's. Then, following World War II, Starr (1948) observed that upper-level troughs and ridges whose axes possessed a general northeast-southwest orientation were of the proper shape to convey angular momentum northward. In Figure 5 a ridge of this sort extends northeastward from the high-pressure center marked with an "X" in the lower right portion of the map. To the east of this ridge, the isobars intersect the 45th parallel nearly at right angles; the air therefore crosses the parallel from almost due north, and carries no angular momentum with it, except that which it possesses as a result of rotating with the earth. To the east of the ridge, the air crosses from the southwest rather than the south, and therefore carries considerable additional angular momentum. The result of this exchange of air across the 45th parallel is therefore a net removal of angular momentum from the south side to the north. Starr felt that throughout middle latitudes in the northern hemisphere, troughs and ridges of this shape were more prevalent than their mirror images, which would transport angular momentum in the opposite direction.

By 1950 routine upper-level wind observations in the northern hemisphere were of sufficient quantity and quality to put Starr's ideas to test. The calculations clearly indicated that the secondary systems, and particularly the upper-level troughs and ridges, transported enough angular momentum across the 30th parallel to maintain the prevailing westerlies north of there against the dissipative effects of friction (Starr and White, 1951). A decade later southern hemisphere observations were plentiful enough to yield a similar result (Obasi, 1963). Thus Hadley's account of the circulation was finally overthrown, not because of any fatal error in his reasoning, but because it failed to agree with observations which after more than two centuries had finally become available.

Close to the equator Hadley's ideas fared better. Air does appear to rise near the equator, notably in the inter-tropical convergence zone, and move toward the poles aloft, but it generally sinks and returns equatorward while still in the subtropics. The resulting closed circuits, which are confined mainly to the equatormost thirty degrees of either hemisphere, and which can convey significant amounts of angular momentum across the 15th parallels, are now known as the Hadley cells. Across the 30th parallels, and into the regions of the prevailing westerlies, the required momentum transport is accomplished by other means—the secondary systems.

Does this revised picture of the circulation explain why the trade winds and the prevailing westerlies exist, and why they are found in their observed locations? I feel that it does not, even though it reveals the immediate cause. We have simply replaced the problem of explaining these currents by the equally formidable problem of explaining why the upper-level troughs and ridges assume the orientations which they do, rather than essentially north-south orientations without much transport of momentum. Before attacking this problem we must consider a more basic question: why do we have secondary circulations at all?

There are many contributing factors. For one thing, the oceans and continents and the mountains and plains are rather irregularly distributed over the earth, and any circulation temporarily showing no variations with longitude could not maintain such a condition. Nevertheless, theoretical studies aimed at determining the effect of the irregularities of the earth's surface indicate that the variations with longitude which they demand are far less pronounced than the variations actually observed. There should therefore be some other explanation.

Although some meteorologists (Eady, 1950) had previously suggested that the secondary circulations formed as a result of the instability of the circulation which would otherwise prevail, I feel that the meteorological world was first made aware of the significance of instability by a laboratory device best known as the "dishpan." The relevant experiments were performed at the University of Chicago in the early 1950's (Fultz, et al., 1959). Although the complete apparatus was rather elaborate and expensive, one of the principal elements was an ordinary dishpan. This was placed on a rotating turntable and filled to a depth of a few centimeters with water. The pan was heated near its rim by a heating coil, and in some cases was cooled near its center by a spray of water from below. The dispan was supposed to simulate a hemisphere of the earth, the heating and cooling simulated the heating and cooling of the atmosphere in equatorial and polar regions, the rotation simulated the rotation of the earth, and it was hoped that the resulting circulation in the dishpan would simulate the circulation of the atmosphere.

One feature of the atmosphere prominently lacking in most of the dishpan experiments was the irregularity of the bottom surface, which would have needed to simulate oceans and continents, or mountains and plains. Within the limits of experimental control, the input was perfectly symmetric with respect to the axis of rotation, and one might have anticipated that the resulting circulation would be symmetric also.

Figure 6 shows a nearly symmetric circulation which developed in one experiment. The photograph is a time exposure of the free surface of the water, upon which particles of a tracer have been sprinkled; the moving particles therefore appear as streaks, and the lengths of the streaks indicate the speed of the flow. The camera rotates with the dishpan, so that only the motion relative to the pan is revealed.

There is a single large vortex, whose center is near the center of the pan. Altogether the flow bears considerable resemblance to the circulation envisioned by Hadley.

Figure 6. Motion at free surface of water in differentially heated dishpan rotating at 1.9 rpm, after statistically steady state has been attained. Radius 19.5 cm, depth of water 4.2 cm. Photo through courtesy of D. Fultz.

Figure 7 shows a circulation obtained in an experiment where the external conditions are the same as before, except that the turntable rotates more rapidly. Here the symmetry is gone, and in addition to one concentrated elliptical vortex there are troughs and ridges bearing considerable resemblance to those found on upper-level weather maps. There is a fairly well-developed jet stream, indicated by the longest streaks, which extends fairly close to the rim at some longitudes.

Sometimes a dye is introduced into the water to reveal the circulation at greater depths. When troughs and ridges occur at the free surface, small vortices resembling the migratory cyclones and anticyclones on weather maps are frequently found below.

Thus a symmetric input sometimes brings about a symmetric flow, and sometimes not. The rather abrupt transition from the type of flow pictured by Hadley to the type of flow more closely resembling the true atmospheric behavior, as the speed of rotation passes some critical value, strongly suggests that the asymmetries, when they occur, are the result of instability. That is, symmetric flow appears to be a mathematical possibility for any rate of rotation, in the sense that it is a solution of the mathematical equations governing the flow. For the higher rates of rotation, however, it appears to be unstable; asymmetric disturbances of small amplitude superposed upon a symmetric flow would ultimately develop into major features of the circulation.

The phenomena of stability and instability play a fundamental role in many of the sciences. The transition from stable to unstable motion is typified by the spinning top, which continues to stand on its point if it spins rapidly, but falls over if it spins too slowly, even though a slowly spinning or even a stationary top standing on its point is a mathematical solution of the equations governing the motion of the top. There is thus an analogy between the spinning top and the rotating dishpan, one of the obvious differences being that whereas the top is unstable when it spins slowly, the dishpan is unstable when it spins rapidly.

How about the real atmosphere? Is the presence of secondary systems an instability phenomenon? We have noted that some asymmetries are to be expected in any case, but that they need not be so pronounced as the asymmetries actually observed. It therefore appears likely that secondary systems of the observed intensity, and particularly the migratory ones, occur because simpler circulation patterns, although mathematically possible, are unstable.

This conclusion has gained further support from one of the most recent innovations in theoretical meteorology—numerical simulation of the circulation, first introduced by Phillips (1956). Numerical simulation is an outgrowth of another recent development—numerical weather prediction. Here one attempts to forecast the weather by solving the system of equations governing the behavior of the atmosphere. We have already noted that some approximations are essential; in particular, because the equations are nonlinear, numerical methods of integration must be used to obtain time-dependent solutions. The initial conditions represent the present weather, and the solution is extended over the range of the forecast—most frequently one or two days.

Figure 7. Motion at free surface of water in dishpan heated as in Figure 6, rotating at 3.8 rpm, after statistically steady state has been attained. Photo through courtesy of D. Fultz.

Figure 8. Northern hemisphere weather map at sea level, generated in numerical experiment by Smagorinsky, *et al.* (1965). Shading is between alternate isobars, additional dashed lines are isotherms. Figure reproduced through courtesy of *Monthly Weather Review.*

In numerical simulation the equations are solved by the same methods, but the initial conditions need not represent any known weather situation, and the solution is extended over a period of months or even years. The numerical solutions are then treated as data, and various statistics are computed from them. The investigator hopes that these statistics will be representative of the general solution of the equations, just as the climatologist hopes that the statistics which he computes from weather records will be representative of the long-term climate.

Figure 8 shows a particular sea-level weather map generated by Smagorinsky (1965) in his numerical experiments. As in the real atmosphere (Fig. 4), there is a nearly unbroken belt of trade winds in the lower latitudes, while the prevailing westerlies occur only in statistical sense, and cyclones and anticyclones are abundant. Evidently the experiment does a creditable job of simulating the atmosphere.

The equations used in numerical simulation may be simplified to the extent of omitting all the inhomogeneities of the earth's surface and suppressing the annual and diurnal variability of the heating. In this case steady-state symmetric solutions may be found numerically. Here there is no need to postulate that the symmetric flow is unstable; its stability may be tested by choosing initial conditions representing the symmetric flow plus a small asymmetric perturbation, and then solving the equations numerically. Within a few days the simulated circulation acquires secondary systems resembling those in the real atmosphere.

I therefore feel that the existence of secondary circulations—the cyclones and anticyclones and the troughs and ridges—has been reasonably well accounted for. How can we now explain the preferred orientation of the trough and ridge lines, as we must if we are to explain the trade winds and the prevailing westerlies? I do not know of any simple qualitative arguments like Hadley's or any simple mathematical demonstrations which accomplish this end.

Nevertheless, in all the major experiments in numerical simulation of the circulation, the trough and ridge lines show a preference for the proper orientations, and the trade winds and the prevailing westerlies appear in the proper latitudes. In a sense, then, these global currents are explained; they are demanded by the system of equations which governs the atmosphere.

Some persons, however, would not find such an explanation very satisfying. They would argue that since the real atmosphere does obey the governing equations, and since trade winds and prevailing westerlies do occur, we know even without examining the equations that they demand the presence of trade winds and prevailing westerlies. To these persons the numerical experiments are little more than a demonstration that we are using realistic equations, and handling them properly.

Yet mathematical solutions do constitute acceptable explanations for many physical phenomena. What is lacking in this instance is a real physical insight into the mechanism through which the troughs and ridges acquire their typical orientation. If there is a simple process which could readily be described in a qualitative manner, it has so far been obscured by the complexity of the total problem.

Having satisfied ourselves reasonably well that the existence of middle-latitude cyclones is the result of instability, can we make the same statement about tropical hurricanes, which in some respects are so similar? It would be natural to assume that hurricanes develop because of the instability of the undisturbed trade winds, but we have yet to demonstrate that this is so. We have not shown that the low-latitude flow is unstable with respect to disturbances having the dimensions of tropical hurricanes, and hurricanes have not appeared in the numerical simulations of the total circulation. In this sense the simulations are less realistic than we should wish. Moreover, we are not sure why hurricanes do not appear.

A very promising recent suggestion (Ooyama, 1962; Charney and Eliassen, 1963) is that we are not properly taking into account the cumulus clouds which are present in the trade wind belts before a hurricane begins to develop. From the macroscopic point view, a mass of atmosphere filled with cumulus clouds is a different fluid from a mass of air which is either entirely saturated or entirely unsaturated with water vapor. In a cumulus-filled atmosphere a slight over-all increase in moisture does not cause a large unsaturated region to become suddenly saturated; it simply increases the percentage of the atmosphere occupied by the clouds, and reduces the spaces between the clouds. Our failure to simulate hurricanes numerically may thus arise because we are trying to make them develop in the wrong fluid.

Likewise, it is not known how great a role hurricanes play in maintaining the currents of larger scale. This lack of knowledge results largely from the lack of enough representative observations in the vicinity of hurricanes. Indeed, meteorologists who have been asked what would aid them most in furthering the purely theoretical study of hurricanes have frequently wished for more detailed observations. Evidently they have not expected to deduce the features which they have not yet observed. If we some day find that hurricanes are instrumental in maintaining the low-latitude circumpolar currents, just as the middle-latitude cyclones are instrumental in maintaining the prevailing westerlies, and if in additon we find that hurricanes cannot be properly explained without taking cumulus convection into account, we shall have established a close interrelation between three widely different scales of motion in tropical latitudes.

One of the most frequent complaints of the meteorologist is that much of the general public thinks of him as primarily a weather forecaster. A glance at such studies as the dishpan experiments or the computation of transports of angular momentum reveals that such a notion is ill founded. Nevertheless, the problem of weather forecasting does occur among the many problems faced by today's meteorologist, and I should like to conclude by saying a few words about it.

I have already mentioned numerical weather prediction. For forecasting in the one-day to one-week range, this powerful method seems destined to become more and more widely used, and it is well to note that certain limitations will nevertheless remain. Even the largest digital computer is a finite instrument, and the current state of the atmosphere must be respresented by a finite collection of numbers, although it may be a large finite collection—perhaps 50,000 numbers. These numbers might be the values of four meteorological parameters—say temperature, two components of the wind, and water vapor content—at each of five elevations at each of 2500 geographical locations. Effectively the machine will then solve a system of 50,000 ordinary differential equations in 50,000 unknowns. Such a system might seem big enough for any physical problem.

But a total of 2500 points means that there is only one point for each 80,000 square miles of the earth's surface. Systems as large as the line of thunderstorms in Figure 1 can easily lie hidden between these points, and may be represented very poorly or not at all.

Computers are becoming ever larger, but it is difficult to imagine one which can adequately describe the structure of every thunderstorm. It is equally difficult to imagine an observational network which will record the structure of every thunderstorm, even if the computer were large enough. Thus there will always be some uncertainty in the initial state.

We have noted that certain simple solutions of the governing equations are unstable. However, numerical simulations and theoretical considerations both indicate that all solutions of the equations are unstable; that is, two solutions originating from slightly different initial states will eventually evolve into considerably different states (Lorenz, 1963).

Now the state of the atmosphere as it is observed and recorded, and the state as it actually exists, may be regarded as two slightly different initial states. The predicted behavior and the acutal behavior will therefore diverge from one another. Thus no method of forecasting can be expected to produce good forecasts for the far distant future.

The numerical simulations further indicate that small differences between the predicted and actual distributions of the weather elements may double in about five days, in the root-mean-square sense. This figure is highly tentative, but if it is correct we should some day be able to forecast a week in advance as well as we now forecast one or two days in advance. Forecasting the general trend of the weather may be possible at much longer range; for example, we may be able to say whether next summer will be a warm summer or a cold one. It seems most unlikely, however, that we shall ever make good weather forecasts for a particular day a month or more in advance.

References

Charney, J. G., and Eliassen, A., 1963: On the growth of the hurricane depression. *Journ. Atmos. Sci., 21,* 68–75.

Coriolis, G., 1835: Sur les équations du mouvement relatif des systèmes de corps. *Journ. Ecole Polytechn., 15,* 142–154.

Eady, E. T., 1950: The cause of the general circulation of the atmosphere. *Centenary Proc. Roy. Meteor. Soc.,* 156–172.

Fultz, D., Long, R. R., Owens, G. V., Bohan, W., Kaylor, R., and Weil, J., 1959: *Studies of thermal convection in a rotating cylinder with some implications for large-scale atmospheric motions.* Meteor. Monographs, Amer. Meteor. Soc., Boston, 104 pp.

Hadley, G., 1735: Concerning the cause of the general trade-winds. *Phil. Trans. Roy. Soc., 39,* 58–62.

Jeffreys, H., 1926: On the dynamics of geostrophic winds. Quart. Journ. Roy. Meteor. Soc., *52,* 85–104.

Lorenz, E. N., 1963: The predictability of hydrodynamic flow. *Trans. New York Acad. Sci., Ser. 2, 25,* 409–432.

Obasi, G. O. P., 1963: Poleward flux of atmospheric angular momentum in the southern hemisphere. *Journ. Atmos. Sci., 20,* 516–528.

Oberbeck, A., 1888. Über die Bewegungserscheinungen in der Atmosphäre. *Sitzungsberichte Königl. Preus. Akad. Wiss.,* 383–395 and 1129–1138.

Ooyama, K., 1962: A dynamical model for the study of tropical cyclone development. *Bull. Amer. Meteor. Soc., 43,* 666.

Phillips, N. A., 1956: The general circulation of the atmosphere: a numerical experiment. *Quart. Journ. Roy. Meteor. Soc., 82,* 123–164.

Smagorinsky, J., Manabe, S., and Holloway, J. L., 1965: Numerical results from a nine-level general circulation model of the atmosphere. *Mon. Weather Rev., 93,* 727–768.

Starr, V. P., 1948. An essay on the general circulation of the earth's atmosphere. *Journ. Meteor., 5,* 39–43.

Starr, V. P., and White, R. M., 1951: A hemispherical study of the atmospheric angular-momentum balance. *Quart. Journ. Roy. Meteor. Soc., 77,* 215–225.

Robert M. Garrels
Abraham Lerman
Fred T. Mackenzie

Controls of Atmospheric O_2 and CO_2: Past, Present, and Future

Geochemical models of the earth's surface environment, focusing on O_2 and CO_2 cycles, suggest that a dynamic steady-state system exists, maintained over time by effective feedback mechanisms

During the past 600 million years (Phanerozoic time), the earth has supported a complex population of organisms. Their numbers and diversity may have waxed and waned, but the ocean and the atmosphere remained within the ranges of compositions necessary for the survival and evolutionary change of life (e.g. Margulis and Lovelock 1974). Because all the carbon dioxide of the atmosphere is used by photosynthesizing organisms and enters and leaves suface ocean waters every few years, and because it is all used in weathering processes on the land every few thousands of years, efficient feedback mechanisms must have operated to hold atmospheric CO_2

All three authors are currently professors of geological sciences at Northwestern University. Robert M. Garrels formerly taught at Harvard University, Scripps Institution of Oceanography, and the University of Hawaii, and spent several years with the U.S. Geological Survey. He has published a number of books on mineral equilibria and chemical cycles. Before joining the faculty at Northwestern, Abraham Lerman taught at the Johns Hopkins University and the Weizmann Institute of Science. He has also worked as a research scientist for the Canada Centre for Inland Waters, and he has done environmental consulting for a number of agencies. His current research focuses on the dynamics of geochemical processes in aquatic and sedimentary environments. Fred T. Mackenzie, Chairman of the Department of Geological Sciences at Northwestern, has taught and done research at a number of universities and research groups, including Harvard University, the University of Hawaii, Bermuda Biological Station, and Shell Oil Company. His current work is centered on interdisciplinary approaches to environmental problems that relate to the past, present, and future of the earth's surface. Garrels and Mackenzie have collaborated on two books related to the subject of this article: Evolution of Sedimentary Rocks *(1971) and* Chemical Cycles and the Global Environment *(with C. A. Hunt; 1975). Address: Department of Geological Sciences, Northwestern University, Evanston, IL 60201.*

within relatively narrow limits during all the environmental changes of Phanerozoic time. If at any interval during that vast time span CO_2 had dropped to less than one third of its present value, almost all photosynthesis would have stopped. The fossil record tells us that such an event has not occurred.

The argument for maintenance of atmospheric CO_2 within a narrow concentration range, despite its circulation in and out of the atmosphere millions of times, can be extended to other components of the earth's surface system. Chemical elements such as calcium, magnesium, sodium, potassium, silicon, sulfur, and carbon have circulated many, many times from land to sea and back again, from land or sea to the atmosphere, or from the sea into the sediments of the ocean floor, to be returned to the land by uplift of the sediments or retreat of the sea. The conclusion that the long-term circulation of materials of the earth's surface environment can be regarded as a dynamic system, protected from severe perturbations by effective feedback mechanisms and without major secular trends, seems reasonable.

In this article we will examine first the distribution of sedimentary rock masses as a function of their ages to draw conclusions concerning the cycling rates of the components of the earth's surface system (the *exogenic* cycle). Then the Phanerozoic system will be modeled in terms of reservoirs and fluxes of major components of the earth's metabolic systems, and we will illustrate some of the feedback mechanisms that seem to have operated to resist or restore perturbations of the system. Most attention will be given to CO_2 and O_2, although it can

be shown that similar feedback mechanisms operate for other chemicals.

The life cycle of sedimentary rocks

Sedimentary rocks are born by deposition from wind and water; they die when they are eroded or transformed chemically into other kinds of rocks. Rock masses can be assigned birth rates and death rates, and they can be subdivided into age groups. The colored portion of Figure 1 shows the observed mass-age distribution of the sedimentary rock population for Phanerozoic time (Gregor 1970). Mass and age data for older sedimentary rocks are insufficient to show the distribution earlier than 700 million years ago, although a complete representation would extend back to at least 3.5 billion years. Even for Phanerozoic time, the smallest age groups that can be separated are 50-million-year intervals. However, it is readily apparent from Figure 1 that the mass diminishes in an apparently exponential fashion from the present back to about 350 million years ago; after an increase, it then again decreases exponentially back to 700 million years.

This relation has been noted by several investigators (c.f. Gregor 1970; Li 1972; Garrels and Mackenzie 1969, 1971a,b) and is subject to different interpretations. Here we present ours, which is consistent with the data and with a dynamic Phanerozoic "average" circulating exogenic system. Figure 1 also shows a calculated sedimentary rock distribution according to our model, superimposed on the observed distribution. We have extended the model back to 2 billion years, but that part of the model beyond 700 million years is entirely speculative in terms of the times of the maxima and minima shown.

The model (essentially that of Li 1972 or Garrels and Mackenzie 1971a,b) assumes a constant rate of sediment birth, a constant *total* death rate, and, of course, a constant total mass. Despite the overall constancy, the model accommodates variations in specific rates, and in fact we can discern two subcycles having distinctly different death rates. The model shows that young sediments are more susceptible to destruction than older ones, and that their mass decays exponentially at a relatively high rate. If this rate had been maintained since the beginning of sediment deposition, it would have left almost no record of very old rocks. However, about 350–400 million years ago, something happened to slow the rate of destruction of the sedimentary rocks then in existence, thus accounting for the presence today of the maximum in mass of sedimentary rock of that age. Perhaps this event was the folding and distortion of the sediments. As a result of such folding, areas of outcrop relative to mass would have been diminished, the probability of destruction of mass declined, and the half-life of these older sediments increased. After that event a new subcycle began, during which the deposition of sediments has been proportional to the mass of the existing sediments, with a constant death rate of 22% per million years, or a half-life of about 150 million years. In the model the death rate of all sediments older than those of the current subcycle is assumed to be the same at any instant of time and is adjusted to maintain the constancy of the total sedimentary rock population.

A test of the model, which was initially developed to account for the age distribution of sedimentary *masses*, has recently become available. Blatt and Jones (1975) determined the relative areas of exposure to erosion of sedimentary rocks as a function of their ages. Figure 2 compares their relative areas, as a function of 100-million-year age groups, with the predicted erosional rate of the same age groups from our model. The correspondence between percentage of area exposed and percentage contribution to present-day deposition is obvious and well within the probable error of mass and area estimations.

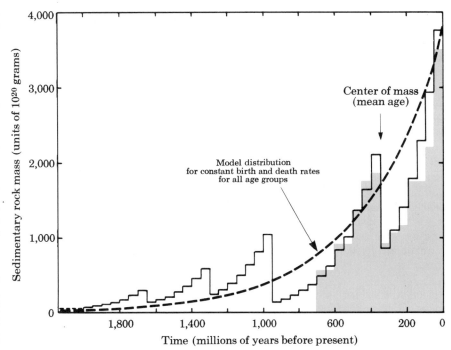

Figure 1. The age distribution of earth's sedimentary rock mass is shown both as estimated from direct measurement (*color*) and as calculated from the model (*black outline*). As the graph shows, the model is consistent with the data and with the concept of a dynamic average circulating exogenic system throughout Phanerozoic time. Values are for age groups at 50-million-year intervals. The model is based on the following conditions: (1) Constant birth rate of sediments of $3{,}750 \times 10^{20}$ grams per 50×10^6 years; (2) constant total death rate of sediments by erosion or metamorphism; (3) constant total mass of sediments of $32{,}000 \times 10^{20}$ grams; (4) half-life of most recent cycle equal to 150×10^6 years; (5) half-life of older cycles adjusted to maintain constant mass, but the same for all older cycle rocks at any time; (6) average life expectancy of rocks of 350×10^6 years.

The present-day age distribution of the population of sedimentary rocks is thus consistent with the assumption of a nearly constant total rate of deposition and destruction throughout Phanerozoic time. It must be emphasized that the sampling has been limited to large age groups, and any fluctuations with response times of less than 50 million years or so would not be seen. Even so, the sedimentary rock population distribution is in accord with the concept of average steady-state cycling of sedimentary mass during Phanerozoic time. As a corollary, to a first approximation, steady-state cycling of mass can be shown to be in accord with a model in which the composition of materials reaching the ocean via streams, glaciers, wind, and groundwater flow has been nearly constant over Phanerozoic time, suggesting that there has been little *evolutionary* change in seawater composition during the Phanerozoic (Garrels and Mackenzie 1974; Mackenzie 1975). Furthermore, although the average rate of erosion deduced from the model is about 75×10^{14} g/yr, about one third of the

present rate (Garrels and Mackenzie 1971a,b), this is about the rate expected for Phanerozoic time, if man's influence has greatly accelerated erosional rates (Judson 1968) and if continental area today is about 20% greater than the average for Phanerozoic time (Sloss and Speed 1974).

A model for the Phanerozoic system

In a "perfect" dynamic steady-state system, each of the chemical components would circulate in the exogenic cycle at a constant rate, and a network of reservoirs and fluxes could be constructed that would describe the behavior of any given component at any given time. Several kinds of information reveal that the flux rates and reservoir sizes of major components of Phanerozoic exogenic cycles tended to fluctuate markedly. Some examples of the types of information available are presented below.

Organic content and carbon isotopes. The age variation in organic content of Phanerozoic sedimentary rocks

(Ronov 1958; Trask and Patnode 1942) and the carbon isotopic values (δC) of carbonate rocks (Keith and Weber 1964) are shown in Figure 3. The data, which are averages for the various Periods, represent time spans ranging from 30 to 100 million years. Organic carbon values fluctuate between about 0.2 and 0.8 percent by weight, suggesting marked differences in the rate of deposition of organic materials if the total depositional rate of sediments remained nearly constant. On the other hand, there is no definite time trend—a finding that is in harmony with the concept of a dynamic average steady state.

Period values for δC in carbonate rocks show changes of about ±1‰ from an unweighted average of +0.6‰, with no trend. Schidlowski and his colleagues (1975) have demonstrated that the average Phanerozoic δC value applies to Precambrian sedimentary rocks as well. Variations of δC in carbonate rocks are usually interpreted as the result of variations in the ratio of carbon deposited as organic carbon to that deposited inorganically as carbonate minerals. The average value of about +0.6% corresponds to a ratio of total carbon in organics to total carbon in carbonate rocks of about 1:4. A fluctuation of 1‰ from the mean indicates a range in depositional ratio of 1:3.5 to 1:6 (Broecker 1970a)—a relatively minor fluctuation considering the length of Phanerozoic time.

Interestingly, the fluctuations of organic content and δC values as shown in Figure 3 are opposite to the expected pattern. Although most models predict that these two parameters should vary sympathetically (see Junge et al. 1975), organic content is found to rise when δC is low. Whatever the explanation of the data, they are consistent with a negative feedback model, in which any perturbation of the system that produces a change in the ratio of organic to inorganic carbon deposited is compensated, and the system remains nearly constant through time.

Sulfur isotopes. Variations in the sulfur isotopic values (δS) of evaporite deposits, which measure within about 1‰ the sulfur isotopic composition of the seawater from which they precipitated, are large and irregular (Fig. 4). The variations are so large, and the

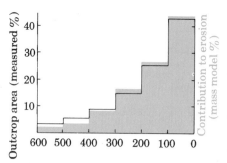

Figure 2. Comparison of the measured area of outcrop of sedimentary rocks (*black outline*) and their predicted contribution to erosion (*color*) over the past 600 million years reveals a correspondence between the two factors that provides support for the model. Age groups are given at 100-million-year intervals.

restoration times from high and low values so long, that it is difficult to decide whether or not there is a dynamic mean that tends to be restored by feedback mechanisms. Low values of δS can be correlated well with times of "excess" gypsum deposition, as shown in Figure 4. According to Garrels and Perry (1974), the range of δS from +30‰ to +10‰ corresponds to a variation of ±30% in the average total amount of sulfate stored in sedimentary rocks and in the ocean. Consequently the evidence from sulfur isotopes tells the same general story that is deduced from secular variation of carbon isotopes and carbon content of rocks: there has been large-scale transfer of sulfate into and out of its reservoir in sedimentary rocks, but mechanisms exist to restore the changes.

Reservoirs and fluxes. We have constructed a system of fluxes and reservoirs of some of the major components of the exogenic cycle involving atmospheric CO_2 and O_2 (Fig.5; Tables 1 and 2). (The reservoirs and fluxes have been taken almost entirely from the work of Garrels and Mackenzie and co-workers; see references.) A similar model has been constructed by Schidlowski et al. (1975); agreement between the two models is quite good and is particularly heartening because they were constructed independently. The chief difference is that Schidlowski and his colleagues used present-day erosional rates to obtain their mean fluxes, whereas we use values about one-third as large, drawing on the model of sedimentary rock mass-age relations (Fig. 1) to obtain erosional rates.

Other authors, for example Berkner and Marshall (1965), Holland (1973), and Walker (1974), have analyzed portions of the exogenic cycle involving O_2 and CO_2.

The model is supposed to represent the mean of Phanerozoic conditions and is shown as a balanced steady-state system. The reservoirs chosen are those considered to participate in major control of atmospheric oxygen and carbon dioxide. We should emphasize that, to include all the important aspects of operation of the exogenic cycle, we would need many more reservoirs and fluxes than are shown in Figure 5. Currently we are wrestling with a model that has about twice as many reservoirs, and we hope someday to be able to simulate the real world with a model sufficiently complex that the results obtained from it will have some valid predictive value. On the other hand, the system shown is sufficient to illustrate the basic behavior of a more inclusive model, and the results obtained from this restricted system can be used to define its inadequacies and to point to the important relations that must be investigated to improve it.

We must use numbers in our model, because an entirely qualitative discussion is impossible. Also, we must discuss the diagram as if there were no questions concerning the accuracy of the numbers or the validity of the processes hypothesized. Otherwise we would require many more pages of justification, qualification, and explanation.

For the sake of discussion, we have given the various reservoirs in the system the numbers 1–7. Fluxes are designated by the letter "F" with subscript numbers representing the two reservoirs involved in the transfer. The direction of flow is indicated by the order of the subscript numbers. For example, F_{23} represents the carbon fixed by photosynthesis which circulates from the CO_2 reservoir (2) to the oceanic biomass (3). At the top center of the diagram is the rapidly operating photosynthetic cycle involving the atmosphere and ocean. Photosynthesis on the land has been omitted, and we have treated it as a closed cycle for which photosynthesis is equaled by oxidation and decay. This is equivalent to saying that significant "leakage" from the photosynthetic cycle—organic material deposited and preserved in sedimentary rocks—is material derived from marine photosynthesis, an assumption in agreement with observation.

The rate of marine photosynthesis is about $2,500 \times 10^{12}$ moles of CO_2 fixed per year (F_{23}) with production of $2,500 \times 10^{12}$ moles of O_2 (F_{21}). Almost all the organic matter formed is reoxidized in the surface ocean, utilizing almost all the O_2 produced (F_{13}). However, about 3.5×10^{12} moles of organic material sinks to the ocean floor each year (F_{34}), causing a *net* production of 3.5×10^{12} moles of oxygen in the atmosphere. Of the material that sinks to the ocean floor (F_{34}), about 30% is oxidized to CO_2 (F_{42}) by bacterial reduction of sulfate to iron sulfide (pyrite) (F_{46}), and returns to the ocean and atmosphere. The remaining 70% is preserved in the sediments (F_{45}).

Of the 3.5×10^{12} moles of oxygen added to the atmosphere each year by photosynthesis, 30% is used in oxidizing reduced substances in rocks exposed to erosion (F_{16}). Here we have called all this reduced material pyrite. A more complete model would have a separate reservoir for other reduced iron compounds, but the amount of oxygen they use is small in comparison to that used in the oxidation of pyrite. The size of the reduced substances reservoir (Res. 6) is given in moles of oxygen demand—the moles of oxygen required to oxidize the entire reservoir to sulfate and ferric oxide. The remaining 70% of the net production of oxygen in the atmosphere by photosynthesis is assigned to the oxidation of organic materials in rocks that have been exposed to erosion at the land surface (F_{15}). It is worth noting, despite our disclaimer on explanation and justification, that present-day estimates of the amount of old organic carbon oxidized (F_{15}) are in good agreement with estimates of new organic carbon buried (F_{45}). The CO_2 resulting from oxidation of old organics returns to the atmosphere (F_{52}). Thus the dynamic balance of atmospheric oxygen during Phanerozoic time rests on the condition that *new organics reaching the seafloor each year equal pyrite and old organic material oxidized during weathering and erosion.*

The $CaCO_3$ reservoir and its fluxes can be regarded as the atmospheric

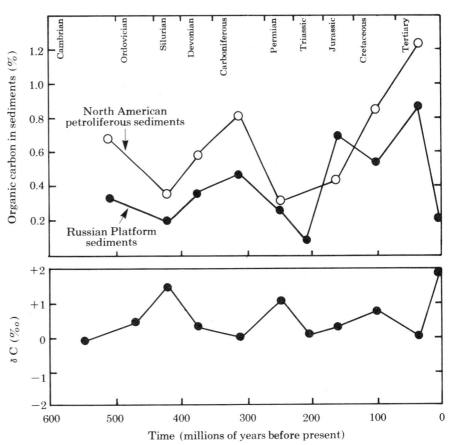

Figure 3. The organic carbon content of sediments and the carbon isotopic composition of carbonate rocks are plotted as a function of geologic age. Values are Period averages. Although the values fluctuate, perturbations that produce a change in the ratio of organic to inorganic carbon are compensated, and the system remains nearly constant through time.

CO_2 buffering system. Carbonate rocks are dissolved by reaction with water and CO_2 (F_{27}), and the products are carried to the ocean by streams (F_{74}); there $CaCO_3$ is precipitated, returning CO_2 to the atmosphere (F_{42}). The weathering of rocks containing calcium silicates (not represented here) also uses CO_2 from the atmosphere. The calcium entering the ocean via this route precipitates from the ocean as $CaCO_3$, and a return of CO_2 to the atmosphere by the conversion of $CaCO_3$ to $CaSiO_3$ at depth in the earth is required to keep the system in balance.

The fluxes into the ocean from the weathering of $CaCO_3$ are much larger than those for $CaSiO_3$, however, and the ability of the system to act as a buffer for atmospheric CO_2 derives from the maintenance of an approximate equilibrium between CO_2 in the atmosphere and an ocean saturated with $CaCO_3$. Dissolved CO_2 in the oceans (including CO_2 and HCO_3^-) is 60 times more abundant than atmospheric CO_2, and additions of CO_2 to the atmosphere tend, over a few

thousand years, to redistribute in approximately the same ratio. If CO_2 is drawn from the atmosphere by processes involving photosynthesis and organic carbon burial (the network in the center of Fig. 5), the ocean tends to give up its CO_2 to restore the atmosphere.

In our model the assumption is made that the oceans are continuously in equilibrium with calcium carbonate and atmospheric CO_2: the only *net fluxes* of CO_2 between the ocean and the $CaCO_3$ reservoir are those required to maintain this equilibrium.

Feedbacks. The operation of the Phanerozoic system and its feedbacks can be illustrated by first following the qualitative effects of a perturbation of the steady-state system. For example, man has increased erosional rates by deforestation, overgrazing, and other activities. Increased erosion exposes more pyrite and old organic carbon to oxidation, increasing the fluxes F_{16}, F_{52}, F_{15}, and F_{42} and resulting in a net drain on atmospheric oxygen and an increase in atmo-

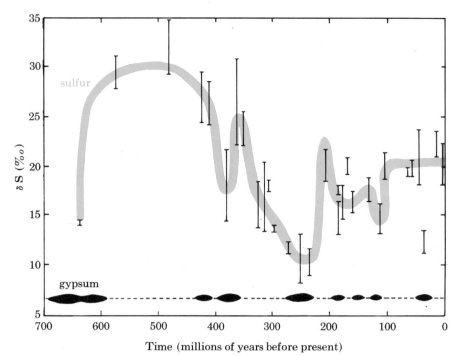

Figure 4. The sulfur isotopic composition of evaporites is shown as a function of geologic age. Colored curve represents the generalized trend of deposition over time; bars bracket the range of values. Despite large fluctuations, the system as a whole shows no major secular trend.

spheric CO_2. Because the rate of photosynthesis in the steady-state system is assumed to be controlled by nutrients such as phosphorus and nitrogen rather than CO_2, presumably there would be little change in either photosynthetic rate or oceanic biomass as a result of increased atmospheric CO_2, and the CO_2, over a period of thousands of years, would redistribute itself between atmosphere and ocean. As oxygen slowly declined, the fraction of the biomass reaching the seafloor would increase, and the net production of oxygen from photosynthesis would increase—a negative feedback to the perturbation. Increased organic material reaching the seafloor (F_{34}) would increase the amount of sulfate reduced (F_{46}) and increase the flux of reduced materials to Reservoir 6. This negative feedback mechanism would counteract the decrease in size of the reduced reservoir as a result of the accelerated rate of erosion.

Similarly, the increased storage of organic material in the organic reservoir (F_{45}) would tend to offset the decrease in size of that reservoir caused by accelerated erosion. Solely on the basis of those qualitative chains of responses we can conclude that atmospheric oxygen, if the initial increased rate of erosion were main-

tained, might well come to a new lower steady-state value, CO_2 to a somewhat higher value, and the reduced materials and organic reservoirs would be diminished. The time scale for changing to the new steady state would be quite long, because the changes in the fluxes are of the order of perhaps 1×10^{13} moles/year, whereas most reservoirs involved are of the order of 10^{20} moles. In 100,000 years, for example, the reservoirs (except CO_2) would change by only a percent or so.

Numerical modeling

For a system of fluxes and reservoirs of even the modest complexity of the one shown in Figure 5, quantitative predictions of the results of perturbations require functional relations between reservoirs and fluxes. We have attempted to set up such functions, and the ones we have chosen are given in Table 3. We are the first to admit that these relations are not at all well defined; but we suggest that it is of greatest importance, if the responses of the earth's metabolism to perturbations are to be understood, to do experiments and make observations that will define them. Also, we emphasize that our whole treatment of the problem is geared to long-term responses, of the order of millions of

years or more, and that prediction of short-term responses requires subdivision of the reservoirs and fluxes into much smaller sub-reservoirs with their accompanying fluxes.

The functional relationships listed in Table 3 are derived chiefly from observations of the behavior of the materials of the various reservoirs under present-day conditions, plus what we regard as reasonable deductions, plus a few guesses. A brief discussion of the flux-reservoir relations, in the order listed in Table 3, may help to show the criteria we have used.

The first relation given—correlating fraction of organic material reaching the ocean floor with the mass of oxygen in the atmosphere—is probably the most important relation of all. It has been derived in an indirect way. According to Schidlowski et al. (1975) and in agreement with our own data, the percentage of organic material in sedimentary rocks shows no trend with time, even if the time range examined is 3.5 billion years. The remarkable conclusion, assuming roughly constant rates of sedimentary rock deposition, is that the functional relation between amount of organics preserved per year and mass of oxygen in the atmosphere must be one that is nearly independent of oxygen mass.

In the absence of free oxygen, with fermentative decay being the principal decay mechanism, the fraction of photosynthesized material that became buried in the oceans must have been substantial. Under present conditions, the fraction of photosynthesized organics that become buried in sediments is only about 0.1%. Thus we are presented with the situation illustrated in Figure 6. As atmospheric oxygen increases, the rate of marine photosynthesis increases and the oceanic biomass increases proportionately, but the flux of organic material to the seafloor remains essentially constant. A doubling of the biomass, at constant oxygen, would double the flux of organics to the seafloor because the fraction oxidized would remain the same. This is the basis for the reservoir-flux relation given as No. 1 in Table 3.

Flux relations Nos. 2 and 3 are based on the observation that no pyrite and no second-cycle eroded organic materials are deposited in new sedi-

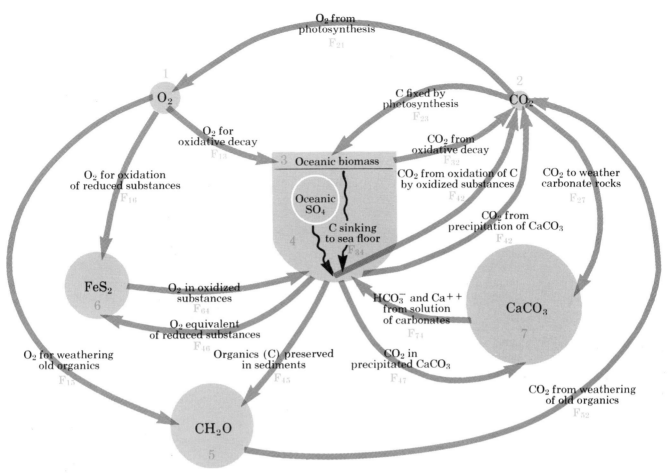

Figure 5. The diagram represents the relationships among the reservoirs and fluxes for the major components of the exogenic cycle involving CO_2 and O_2. See Tables 1 and 2 for magnitudes of reservoir masses and values of fluxes.

ments. Also, the minimum oxygen pressure required for oxidation of pyrite or organic materials is extremely low. Furthermore, detrital or second-cycle pyrite has not been found in significant quantities in sedimentary rocks of marine origin younger than about 2×10^9 years. Even in rocks older than 2×10^9 years most pyrite seems to be the result of reduction of sulfate, undoubtedly by bacteria. Consequently the oxygen demand of the reduced reservoir was made proportional to the size of the reservoir, that is, proportional to the rate of erosion of pyrite-bearing rocks.

The relation of the mass of free oxygen in the atmosphere to the rate of oxidation of old organic materials exposed to erosion is assumed to be the same as that for pyrite. The basic argument for making the rate of oxidation dependent only on the size of the organic reservoir—that is to say, on the rate of exposure of old carbon to the atmosphere—is the lack of a trend in the percentage of organic

Table 1. Sizes of reservoirs of major components of the exogenic cycle involving CO_2 and O_2 (units of 10^{18} moles)

C in living biomass (Res. 3)	0.0006
CO_2 in atmosphere (Res. 2)	0.055
C in ocean as CO_2 and HCO_3^- (Res. 4)	3.234
Ca in ocean (Res. 4)	14
O_2 in atmosphere (Res. 1)	38
O_2 in $SO_4^=$ in ocean (Res. 4)	84
O_2 total demand for substances (primarily FeS_2) in reduced reservoir (Res. 6)	400
C in sedimentary organics (Res. 5)	1,000
$CaCO_3$ (Res. 7)	4,000

Table 2. Sizes of fluxes of major components of the exogenic cycle involving CO_2 and O_2 (units of 10^{12} moles/year)

O_2 from photosynthesis (F_{21})	2,500
C fixed by photosynthesis (F_{23})	2,500
O_2 for oxidative decay (F_{13})	2,496.5
CO_2 from oxidative decay (F_{32})	2,496.5
O_2 for oxidation of reduced substances (F_{16})	1
O_2 for weathering of old organics (F_{15})	2.5
O_2 in oxidized substances (F_{64})	1
O_2 equivalent of reduced substances (F_{46})	1
C sinking to sea floor (F_{34})	3.5
C preserved in sediments (F_{45})	2.5
CO_2 from oxidation of C by oxidized substances (F_{42})	1
CO_2 from weathering of old organics (F_{52})	2.5

Table 3. Functional relations among fluxes and reservoirs of the Phanerozoic system

1. $\log_{10} F_{34(C)} - \log_{10} F_{23(C)} = -k_1 M_{1(O_2)}$ $k_1 = 7.5102 \times 10^{-20}/yr$

2. $F_{16(O_2)} = k_{16} M_6 \text{(oxygen demand)}$ $k_{16} = 2.5 \times 10^{-9}/yr$

3. $F_{15(O_2)} = k_{15} M_{5(C)}$ $k_{15} = 2.5 \times 10^{-9}/yr$

4. $F_{64(O_2)} = F_{16(O_2)}$

5. $F_{52(O_2)} = F_{15(O_2)}$

6. $F_{46} \text{(reduced substances)} = k_{46} F_{34(C)} M_{4(O_2 \text{ in } SO_4^-)}$ $k_{46} = 3.40 \times 10^{-21}/\text{mole}$

7. $F_{45(C)} = F_{34(C)} - F_{46} \text{(reduced substances)}$

8. $M_{2(CO_2)} = k_2 M_{4(Ca)} M_4^2 {}_{(HCO_3)}$ $k_2 = 3.78 \times 10^{-40}/\text{mole}^2$

9. $M_{4(CO_2)} = k_3 M_{2(CO_2)}$ $k_3 = 0.254$

10. $F_{21(O_2)} = F_{23(C)}$

11. $F_{13(O_2)} = F_{32(CO_2)}$

12. $F_{13(O_2)} = F_{21(O_2)} - F_{34(C)}$

carbon in rocks with time. If a significant proportion of old organic carbon were recycled, the mass of buried carbon should show a continuous secular increase with time. Thus the implication is that, for Phanerozoic time at least, there was always enough atmospheric oxygen, whatever its fluctuations, to oxidize all old carbon exposed to weathering.

The restorative flux for the reduced reservoir (No. 6), formation of pyrite in sediments, is assumed to be proportional to the sulfate concentration of seawater and to the flux of organic material reaching the seafloor. Consequently we have set up this flux as a function of the product of the mass of seawater sulfate and the organic flux to the seafloor. The residual organic material is added to the organic reservoir (No. 7).

It is assumed that the ocean is always in equilibrium with calcium carbonate (No. 8). Today the surface waters are oversaturated and the deep waters are undersaturated, but various independent estimates of present-day rates of addition of calcium to the ocean and its removal are equivalent. Thus our basic tenet is that higher atmospheric CO_2 would result in more dissolved calcium carbonate, and that equilibrium for the reaction

$$CO_2 + H_2O + CaCO_3 = Ca^{++} + 2HCO_3^-$$

would be approached within a few thousands of years after a CO_2 increase or decrease.

Neither a reservoir for calcium sulfate nor fluxes into and out of the ocean

from this reservoir have been included in our model. Today it appears that there is a flux of $CaSO_4$ into the oceans of about 1×10^{12} moles S/yr, but that there is a much smaller return flux from precipitation of $CaSO_4$. Calcium sulfate is currently accumulating in the ocean. On the other hand, there has been little change in δS of the oceans in the last 100 million years (Fig. 4). Because changes in δS are controlled by changes in the ratio of sulfur entering the ocean from $CaSO_4$ and FeS_2 and the ratio of sulfur leaving the oceans as these two species, we can only conclude for the moment that deposition and nondeposition must alternate in such a way as to have maintained a nearly constant reservoir of gypsum for the last 100 million years, even though the size of the reservoir fluctuated markedly prior to that time. Our treatment of the effect of gypsum is to run the system without considering it, and then to estimate

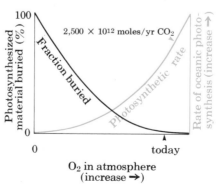

Figure 6. The rate of marine photosynthesis and the fraction of photosynthesized material buried beneath the oceans are shown as functions of the partial pressure of O_2 in the atmosphere.

qualitatively the effects of deposition or nondeposition.

The preceding discussion of the way in which functions have been assigned to the fluxes and reservoirs of Figure 5 may not inspire confidence in the results of our numerical modeling. On the other hand, the results can be tested against what is known about the real system; alternate functions for fluxes and reservoirs can be tested to see if they have a critical influence on the results obtained. We have arbitrarily perturbed the system in various ways, and in the following section we present the effects of these perturbations as obtained from computer calculations. In our opinion, the results demonstrate that a complicated series of negative feedbacks *do* tend to operate in the exogenic cycle, and that time scales for the restoration to a new steady-state system are reasonable. Furthermore, deviations of the new steady-state systems from the chosen "mean system" are within the range permissible for the continuous maintenance of abundant and varied life on earth during Phanerozoic time.

Perturbations of the steady-state system

Increasing the rate of erosion. One of man's effects on the exogenic cycle has been to increase the average rate of erosion since the beginning of Phanerozoic time by a factor of about three. It is difficult to say just when man's influence reached this level. Important anthropogenic effects may date back several thousands of years in the Mediterranean countries and in Africa, where overgrazing, in particular, has been blamed for loss of vegetation, change in albedo of the areas affected, and other important erosional and climatic changes.

In modeling this situation, we increased the global rate of erosion threefold, thus changing the oxygen demand of pyrite and organic carbon exposed to erosion each year by a factor of three. In the model, this corresponds to increasing the oxygen demand by the reduced reservoir and by the organic carbon reservoir, and retaining the assumption that all reduced substances and old organic materials exposed to erosion are oxidized completely. Table 4 shows the output from the computer. The notable changes are that atmospheric

Table 4. Changes in reservoirs and fluxes resulting from tripling the rate of erosion

Time (millions of years)	O_2 reservoir (units of 10^{18} moles)	CO_2 reservoir (units of 10^{18} moles)	Pyrite reservoir (units of 10^{18} moles O_2 equivalent)	Oceanic sulfate reservoir (units of 10^{18} moles O_2 in SO_4)	Organic carbon reservoir (units of 10^{18} moles C)	Carbon flux to seafloor (units of 10^{12} moles/yr)
0	38.0	0.055	400	84.0	1,000	3.5
1	33.3	0.074	399	85.3	997	7.9
2	31.7	0.099	398	85.7	996	9.9
3	31.7	0.115	398	85.8	995	10.4
4	31.7	0.121	398	85.6	995	10.4
5	31.7	0.125	398	85.6	995	10.4
6	31.7	0.128	398	85.5	995	10.4
7	31.7	0.131	398	85.5	995	10.4
8	31.7	0.134	399	85.2	995	10.4
9	31.7	0.137	399	85.4	995	10.4
10	31.7	0.140	399	85.3	995	10.4

Table 5. Changes in reservoirs and fluxes resulting from doubling the rate of photosynthesis and maintaining the increased rate

Time (millions of years)	O_2 reservoir (units of 10^{18} moles)	CO_2 reservoir (units of 10^{18} moles)	Pyrite reservoir (units of 10^{18} moles O_2 equivalent)	Oceanic sulfate reservoir (units of 10^{18} moles O_2 in SO_4)	Organic carbon reservoir (units of 10^{18} moles C)	Carbon flux to seafloor (units of 10^{12} moles/yr)
0	38.0	0.055	400	84.0	1,000	3.5
1	40.2	0.019	401	83.4	1,001	4.8
2	41.1	0.0080	401	83.1	1,002	4.1
3	41.5	0.0042	401	83.0	1,002	3.8
4	41.7	0.0030	401	83.0	1,003	3.7
5	41.9	0.0022	401	82.9	1,003	3.6
6	41.9	0.0020	401	82.9	1,003	3.6
7	42.0	0.0017	401	82.9	1,003	3.5

oxygen reaches a new steady state within about 2 million years at a level about 15% below the present level, whereas atmospheric CO_2 rises to about 2.5 times the present value (~800 ppm). The other reservoirs are not affected greatly. However, the flux of organic carbon to the seafloor is tripled.

The overall response to the system can be summarized as follows: (1) Increase in oxygen consumption by increased erosional rate lowers atmospheric oxygen. (2) Decreased atmospheric oxygen results in a complementary increase in the amount of organic material reaching the ocean floor. (3) The increase in organics reaching the ocean floor increases the return flux of reduced materials to the reduced reservoir (increase in pyrite synthesis) and increases the rate of accumulation of organic materials, to accommodate the increased rate of organic consumption because of the accelerated erosional rate. (4) Atmospheric CO_2 also increases threefold, resulting in increased dissolved $CaCO_3$ in the oceans. (5) The

photosynthetic rate is slightly reduced from $2,500 \times 10^{12}$ moles CO_2/yr to $2,460 \times 10^{12}$ moles CO_2/yr.

Although the rate of deposition of organic materials is increased, the percentage of organic materials in new sediments would remain the same, because of the concomitant increase in other eroded materials. Also, despite some increased storage of $CaCO_3$ in the oceans because of increased atmospheric CO_2, the circulation rate of $CaCO_3$ through the oceans would increase, so that there would be only a slight change in the ratio of organic material buried to inorganic carbonate deposited, and hence little effect on the $\delta^{13}C$ values for organic materials or for carbonates deposited. Furthermore, any isotope effect would be restricted to the few millions of years necessary to reach the new steady state. The increased burial rate of organic material presumably would be supported by an increased nutrient supply from the land to the oceans.

The effects of accelerated erosion,

even though they must be considered over a time scale of several million years, might possibly be greater than those of the burning of fossil fuels, in terms of oxygen depletion and CO_2 increase (Broecker 1970b). Man's greatest long-term contribution to the exogenic cycle may be his drastic effects on erosion rates.

The results of the modeling suggest that an increase in the rate of erosion tends to make the whole machine run faster. Changes in atmospheric oxygen and CO_2 would occur, but a new steady state would develop before the disturbance of previous levels became drastic.

Doubling the rate of photosynthesis. In this second scenario, the perturbation is an instantaneous doubling of the rate of photosynthesis to a constant value. How such a perturbation could be accomplished in the real world is hard to imagine—perhaps it might be a result of man's adding nutrients to the oceans, perhaps a result of increase in the rate at which nutrients well up from deep

Table 6. Doomsday scenario: Changes in reservoirs resulting from cessation of photosynthesis

Time (millions of years)	O_2 reservoir (units of 10^{18} moles)	CO_2 reservoir (units of 10^{18} moles)	Pyrite reservoir (units of 10^{18} moles O_2 equivalent)	Oceanic sulfate reservoir (units of 10^{18} moles O_2 in SO_4)	Organic carbon reservoir (units of 10^{18} moles C)
0	38.0	0.055	400	84	1,000
1	34.5	0.056	399	85	998
2	31.0	0.21	398	86	995
3	27.6	0.42	397	87	993
4	24.1	0.70	396	88	990
5	20.6	1.03	395	89	988
6	17.2	1.41	394	90	985
7	13.8	1.84	393	91	983
8	10.3	2.30	392	92	980
9	7.0	2.80	391	93	978
10	3.6	3.32	390	94	975
11	0.28	3.88	389	95	973
12	Near zero	3.88	389	95	973

oceans. The results of this perturbation are shown in Table 5. According to the model, atmospheric oxygen would climb to a new steady state at a 10% higher value. The feedbacks are as follows: Increase in the rate of photosynthesis would increase the size of the biomass and consequently the flux of organic carbon to the seafloor. More organic carbon would be preserved initially, causing net addition of oxygen to the atmosphere. With a rise in oxygen, the flux of organics to the seafloor would diminish again. The pyrite reservoir would increase slightly, because of the initial increase of organic carbon reaching the seafloor, resulting in an increase in the rate of reduction of oceanic sulfate. Oceanic sulfate would be correspondingly depleted.

However, because of the increase in size of the organic reservoir, CO_2 would be drawn from the atmosphere-ocean system. With loss of CO_2 from the system, the amount of calcium, bicarbonate, and dissolved carbon dioxide in the ocean would diminish, and oceanic pH would rise to about 9. The lowering of atmospheric CO_2 would be so great that it would drop below the level (about 0.15×10^{18} moles) necessary to maintain photosynthesis. In other words, the model shows that an increased rate of photosynthesis could cause CO_2 to become a "limiting nutrient," and a doubled rate of photosynthesis could be maintained only if there were a source of CO_2 not considered by us. It seems likely that this perturbation is an "impossible" one. The simulation is useful simply for

showing that CO_2 supply probably can be an important feedback for controlling the photosynthetic rate, even though it has been shown to be a limiting nutrient today only in a few small freshwater bodies.

Doomsday. We thought it might be interesting to run a Doomsday scenario in which photosynthesis is instantaneously halted. Jacques Cousteau, in a letter from the Cousteau Society (1975), has constructed such a scenario. We quote a few sentences to suggest the kinds of changes and the time scales he envisions: "With life departed, the ocean would become, in effect, one enormous cesspool. Billions of decaying bodies, large and small, would create such an insupportable stench that man would be forced to leave the coastal regions. . . . Then would be visited the final plague, anoxia. . . . And so man would finally die, slowly gasping out his life on some barren hill. He would have survived the oceans for perhaps 30 years."

Table 6 summarizes the results from the model depicting cessation of photosynthesis on the earth. The first event, disappearance of the oceanic biomass within less than a year, takes place so quickly that we cannot show it in the table. Residence time for the biomass is only a few months, so it would be consumed rapidly by oxidative decay. However, its decay would increase atmospheric CO_2 by less than 0.1% and deplete atmospheric O_2 by about 0.001%. Thus we would predict a sterile ocean after a very short time, although there might

be local "cesspool effects" from concentrations of dead organisms near shore.

After that, following Table 6, oxygen would be continuously depleted over a period of 10 million years or so. When it finally disappears, the model becomes inoperative. The time scale shown is probably a minimum, because the rate at which oxygen is consumed by the weathering of pyrite and old organic carbon might well diminish as oxygen pressure became lower. There would be significant changes in the pyrite and oceanic sulfate reservoirs as a result of oxygen consumption by the pyrite reservoir, increasing oceanic sulfate. Without a flux of organic material to the seafloor, oceanic sulfate from pyrite oxidation would accumulate, having no mechanism for reduction, although conceivably it could be removed in $CaSO_4$ deposits. The organic carbon reservoir also would be diminished.

An important point is that, at the time oxygen disappears, the ocean contains even more sulfate than at steady state in an oxygenated world. Reduction of sulfate in the oceans and in the $CaSO_4$ reservoir would have to wait for the transport of old organic carbon into the sea. After free oxygen disappears, old organic carbon could act as a reductant for sulfate. We have not modeled the system, but it would surely take a very long time, perhaps hundreds of millions of years, to remove all the sulfate from the exogenic system.

We agree with Cousteau on the effects

of removing oxygen from the atmosphere, but our time scale gives the human race much more room for contemplation of the end. As modeled, CO_2 pressure would rise by a factor of 70, to about 10^{-2} atmospheres. Oceanic pH would drop to about 7. Undoubtedly the result is fortuitous, but many estimates for the early anoxic earth put CO_2 pressure at about this value. We suspect that feedbacks might come into operation to lessen the rise of CO_2—such as conversion of Mg-silicates to Mg-containing carbonates—like the mineral dolomite.

We have tried to show that even if we use a simple and somewhat incomplete model for the exogenic system, negative feedbacks to perturbations operate to restore the system to new steady states that are not drastically displaced from the original state. The response times for the perturbations we have examined are of the order of millions of years. The only perturbation modeled by us, including several not discussed here, that does not result in a new steady state is that depicting the elimination of life. This is in agreement with Lovelock and Margulis's "Gaia" hypothesis (1974), in that the biosphere plays an active adaptive role in maintaining the earth in a state of equilibrium; in particular, the earth's atmosphere is maintained and regulated by life and its interaction with the total exogenic system.

References

Berkner, Q. V., and L. C. Marshall. 1965. On the origin and rise of oxygen concentration in the Earth's atmosphere. *J. Atmos. Sci.* 2:225–61.

Blatt, H., and R. L. Jones. 1975. Proportions of exposed igneous, metamorphic and sedimentary rocks. *Bull. Geol. Soc. Am.* 86:1085–88.

Broecker, W. S. 1970a. A boundary condition on the evolution of atmospheric oxygen. *J. Geophys. Res.* 75:3553–57.

———. 1970b. Man's oxygen reserves. *Science* 168:1537–38.

Cousteau, J. 1975. Letter of the Cousteau Society, 1975.

Garrels, R. M., and F. T. Mackenzie. 1969. Sedimentary rock types: Relative proportions as a function of geologic time. *Science* 163:570–71.

———. 1971a. Gregor's denudation of the continents. *Nature* 231:382–83.

———. 1971b. *Evolution of Sedimentary Rocks.* NY: Norton.

———. 1972. A quantitative model for the sedimentary rock cycle. *Mar. Chem.* 1:27–41.

———. 1974. Chemical history of the oceans deduced from post-depositional changes in sedimentary rocks. *Studies in Paleo-Oceanography.* Spec. pub. 20, pp. 193–204. Soc. Econ. Paleo. Mineral. Tulsa, OK.

Garrels, R. M., and E. A. Perry, Jr. 1974. Cycling of carbon, sulfur, and oxygen through geologic time. In *The Sea*, ed. E. D. Goldberg, vol. 5, pp. 303–36. NY: Wiley-Interscience.

Gregor, B. 1970. Denudation of the continents. *Nature* 228:273.

Holland, H. D. 1973. Ocean water, nutrients and atmospheric oxygen. In *Proc. I.A.G.C. Mtg.,* Tokyo 1970, 1:68–81.

Judson, S. 1968. Erosion of the land. *Am. Sci.* 56:356–74.

Junge, C. E., M. Schidlowski, R. Eichmann, and H. Pietrek. 1975. Model calculations for the terrestrial carbon cycle: Carbon isotope geochemistry and evolution of photosynthetic oxygen. *J. Geophys. Res.* 80:4542–52.

Keith, M. L., and J. N. Weber. 1964. Carbon and oxygen isotopic composition of selected limestones and fossils. *Geochim. Cosmochim. Acta* 28:1787–1816.

Li, Yuan-Hui. 1972. Geochemical mass balance among lithosphere, hydrosphere, and atmosphere. *Am. J. Sci.* 272:119–37.

Lovelock, J. E., and L. Margulis. 1974. Atmospheric homeostasis by and for the biosphere: The Gaia hypothesis. *Tellus* 26:2–10.

Mackenzie, F. T. 1975. Sedimentary cycling and the evolution of sea water. In *Chemical Oceanography*, ed. J. P. Riley and G. Skirrow, 2nd ed., vol. 1, pp. 309–64. London: Academic Press.

Margulis, L., and J. E. Lovelock. 1974. Biological modulation of the Earth's atmosphere. *Icarus* 21:471–89.

Ronov, A. B. 1958. Organic carbon in sedimentary rocks (in relation to the presence of petroleum). *Geochem.* 5:510–36.

Schidlowski, R., R. Eichmann, and C. E. Junge. 1975. Precambrian sedimentary carbonates: Carbon and oxygen isotope geochemistry and implications for the terrestrial oxygen budget. *Precamb. Res.* 2:1–69.

Sloss, L., and R. C. Speed. 1974. Relationships of cratonic and continental margin tectonic episodes. In *Tectonics and Sedimentation*, ed. W. R. Dickinson, Spec. pub. 22, pp. 98–119. Soc. Econ. Paleo. Mineral. Tulsa, OK.

Trask, P. D., and H. W. Patnode. 1942. *Source Beds of Petroleum.* Menasha, WI: Banta.

Walker, J. C. G. 1974. Stability of atmospheric oxygen. *Am. J. Sci.* 274:193–214.

"Bunsen, I must tell you how excellent your study of chemical spectroscopy is, as is your pioneer work in photochemistry—but what really impresses me is that cute little burner you've come up with."

Chester B. Beaty

The Causes of Glaciation

Major glaciations have been caused not by dramatic changes in the earth's climate, but by the conjunction of several discrete factors at times when the continents were located at high latitudes

Mark Twain is supposed to have said, "The researches of many commentators have already thrown much darkness on this subject, and it is probable that, if they continue, we shall soon know nothing at all about it." The recent outpouring of papers pertaining to possible causes and consequences of glaciation might lead one to believe that the wily Twain, in a remarkable act of prescience, was describing precisely the situation afflicting us today. At this point in time, the student of glaciation will be more handicapped than helped by the quantity of information available, since in all probability he or she will not live long enough to consider it all.

In the face of this explosive proliferation of data and speculation, I offer a relatively simple model of the causes of glaciation, understandable by the generalist but acceptable, in the main, to the specialist. I realize that this may seem a brash undertaking, given the amount of expertise that has been brought to bear on the problem. Nevertheless, by omitting the intricate details, I will develop a model of glaciation incorporating much that is reasonably well known

Chester B. Beaty, Professor of Geography at the University of Lethbridge, Alberta, Canada, received his Ph.D. from the University of California, Berkeley, in 1960. Following 11 years at the University of Montana, he moved to his present position in 1969. His primary research activity has been in geomorphology, with special attention directed to the origin of alluvial fans in the American southwest, and a long-time interest has centered on the relationship between landscape evolution and microclimatic differences. Address: Department of Geography, University of Lethbridge, Lethbridge, Alberta, Canada T1K 3M4.

about ice ages past and present and taking into consideration the variable factors that have controlled atmospheric behavior through time. In this way, perhaps we can shed some light rather than more darkness on the issues.

It should be stated at the outset that most of what follows contains little that is new, original, or startling (see, for example, Emiliani and Geiss 1955; Donn and Shaw 1977). If anything differs markedly from much that has been written and said before, it is my suggestion that we have been relying too heavily upon the concept of major climatic change in seeking the causes of glaciation. I think there can be, and, indeed, is, little direct evidence of climatic change as such. There is abundant evidence from various parts of the earth suggesting changes in kind and intensity of geomorphic processes and in floral and faunal assemblages, but evidence of climatic change *per se* is simply not available. The paleoscientist can thus only *infer* differing conditions from study of what is basically geological, not meteorological, evidence.

What I will argue, then, is that what has changed most in the past has been the *impact* of various climates—all arranged in more or less fixed latitudinal belts—on different parts of the earth's surface. This is not to imply that the behavior of the atmosphere has been invariable through time. On the contrary, there is abundant indirect evidence of climatic fluctuations, some of which are temporarily self-sustaining (in the geologic sense) when once initiated. But do such demonstrable fluctuations constitute climatic *change* of sufficient magnitude, for example, to produce coal on Antarctica or corals in Greenland? It

is worth recalling that one of the great advantages possessed by Alfred Wegener was his realization that on a round earth heated by a single source, the sun, it is easier to move continents (by whatever mechanism) than it is to shift climatic zones. I therefore hold that the gross distribution and intensity of climate have been roughly constant throughout much of geologic history.

Contributing factors

Glaciation represents a particular hydrological response to a specific set of atmospheric conditions, most immediately, conditions in which accumulation of snow exceeds ablation. This simple, fundamental requirement for the initiation of glaciation involves the interrelated functioning of many factors, about which a considerable and ever-expanding literature has built up over the years (see Flint 1957, 1971 for useful summaries). Most investigators would probably accept the following factors as being significant: (1) mountain-building and continental uplift; (2) variation in the characteristics of oceans; (3) variation in solar output; (4) volcanic activity; (5) variation in surface albedo; (6) variations in the earth's orbit; and (7) changes in the latitudes of continents. There may be other, unknown—even unsuspected—factors at play (see, for example, Steiner and Grillmair 1973, in which possible cosmic or galactic causes of glaciation are discussed; and, for perceptive summaries, Mitchell 1968 and Beckinsale 1973), but a usable model of the causes of glaciation can readily be constructed from these seven (Fig. 1).

Mountain formation and continental uplift. It was long ago suggested that

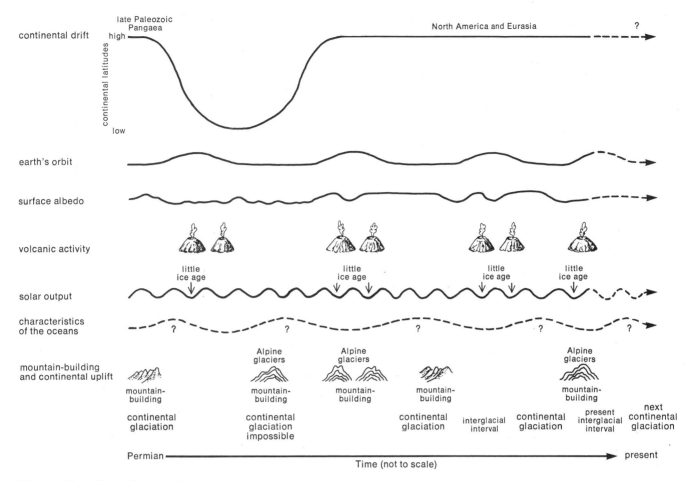

Figure 1. Many discrete factors act in concert to cause major glaciations. In order to stress the conjunctions, the interactions among and variations in these factors through time are deliberately simplified here and are not drawn to scale.

mountain-building might serve as a stimulus to glacial activity. As pointed out by Flint (1957) and others, virtually all of the higher ranges display evidence of multiple late Cenozoic advances and retreats. Even mountains in tropical and subtropical locales supported limited ice masses during the Pleistocene, strongly suggesting that continental elevation has been a factor in the development of glaciation.

But in all probability, high mountains have always existed somewhere on the earth's surface, while massive continental glaciation has been relatively rare. Concomitantly, it is likely that local alpine glaciers have also generally been present, as they are today; given sufficient elevation, mountain glaciers of variable dimensions and duration can generally be expected to appear.

If the formation of mountains has been a factor in the initiation of large-scale glaciation, it has probably been only a coincidental one. The formation of a mountain range in middle to high latitudes athwart the prevailing westerlies, for example, should quickly result in glaciation in the growing highland. Furthermore, especially if such tectonic activity is located on the western side of a mid-latitude continent, it should significantly affect the downwind circulation pattern, tending to generate a stationary trough in the lee of the range and to promote the development of cyclones (and hence perhaps more precipitation) over much of the downwind terrain (Manabe and Terpstra 1974). But it seems evident that the growth of major ranges has not inevitably led to glaciations of a magnitude approaching those of Quaternary, late Paleozoic, and earlier times. And in fact, in North America, which comes to mind as a clear example of the situation just described, the major Cenozoic glaciations did not take place until some

time after the contemporary Cordilleran ranges had been formed. Some factor or factors other than mountain formation must have been involved.

In addition to the potential direct effect of mountain-building itself, the formation of major continental glaciers resulted in a relative increase in land-mass elevations by bringing about a worldwide drop in sea level. Furthermore, the buildup of ice produced an absolute increase of continental altitudes in regions of massive snow accumulation. Estimates of the combined relative and absolute elevation increases have varied. Beckinsale (1973) noted that "this ice-sea vertical-change mechanism assumes a mean maximum magnitude exceeding 1,200 meters ... for polar land masses," and this value is accepted as having at least ball-park accuracy.

Variation in the oceans. Changes in characteristics of the oceans and their circulation patterns have long been suspected of playing a role in short- and possibly long-term variations in atmospheric behavior (for an excellent summary, see Brooks 1970), but the specific nature of the relationship remains to be discovered.

Short-term atmospheric responses to measured changes in surficial oceanic conditions have been identified with some degree of certainty. For instance, the effect on the atmosphere of several years of "anomalously" warm surface waters ("anomalous" in the context of a 40-year record) over the central part of the North Pacific has been described by Namias (1969). He found that from 1961 to 1967, when these warm temperatures occurred, winter temperatures in the eastern two-thirds of the United States averaged perceptibly lower

bearing on the glacial phase of the hydrologic cycle (Namias 1972). But as Lamb says, "Cause and effect, action and reaction between atmosphere and ocean, are intertwined at so many points . . . that they can only be distinguished on this or that time scale." If the time scale is longer than just a few years, confidence in hypothetical relationships must diminish.

An instructive case is Weyl's (1968) "theory of the ice ages," which is based upon the assumption that a decrease in surface salinity of the North Atlantic would lead to an increase in the extent of sea ice, with attendant atmospheric effects, including stimulation of glaciation. Weyl postulated that North Atlantic surface salinity could be decreased by a decrease of water-vapor flux across

creases or decreases in the rate of seafloor spreading *not* readily detectable in the deep-sea core record, might lead to significant alteration of oceanic circulation patterns. This, in turn, could conceivably affect atmospheric circulation in a way favorable to the initiation of glaciation.

Additionally, compartmentalization by continental drift of the great world ocean of late Paleozoic time into three or four more or less discrete basins must have had profound effects on surface and deep-water circulation patterns. But such compartmentalization did not happen suddenly, geologically speaking, whereas initiation of at least some components of the late Cenozoic glaciation apparently did. Since the deep-sea core record is incapable of supplying information about the characteristics of

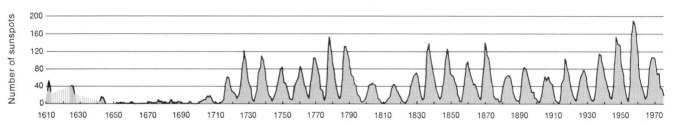

Figure 2. The annual mean sunspot number dropped significantly between about 1645 and 1715. During approximately the same period, world temperatures fell, the snowline lowered, and snow cover in regions of northeastern Canada increased. Early values are, of course, estimates. (From Eddy, Gilman, and Trotter 1977).

than long-term normals and were the result of a mean atmospheric circulation pattern over North America dominated by a persistent ridge along the western side of the continent and a strong trough over the eastern half. Likewise, Lamb (1972) offered a thoughtful discussion of the atmospheric effects of variations in ocean temperatures in the North Atlantic. He showed that atmospheric temperatures, pressure distribution, and circulation patterns in Western Europe clearly respond to changes in sea-surface temperatures, with "anomalously" warm surface water in the North Atlantic producing features in the large-scale atmospheric circulation pattern comparable to those described by Namias for the North Pacific.

Continuation of such climatic regimes for extended periods of time obviously would have significant effects on overall temperature and precipitation trends, with an inevitable

the Isthmus of Panama, brought about by atmospheric circulation changes caused by variations in seasonal insolation due to variations in the earth's orbit. The Weyl model is open to doubt (see Sachs 1976), but even if it is valid, the ultimate *cause* of glaciation would be atmospheric variability, not oceanic behavior, since the decrease in salinity would be the result of an initial atmospheric change. In the Weyl model (and in others), cause and effect indeed become muddled, in spite of ever-increasing knowledge about the oceans.

Accordingly, the oceanic factor is given an equivocal position in the model. Question marks are attached to it in recognition of the fact that our present conception of the influence of air-sea interactions may require a substantial reevaluation in the future. For example, it is possible that subtle but critical changes in the geometry of the ocean basins, produced by in-

the oceans of late Paleozoic time, our ideas about marine environments of that period are of necessity based on speculation and theory.

In recent years, many paleoclimatic hypotheses have been based upon the burgeoning deep-sea core record (Cline and Hays 1976; Savin 1977), but what is represented in that record is, in my judgment, primarily evidence of the *results* of glaciation, not evidence of causes. Accordingly, it is here held that oceanic behavior, at least on the glacial-interglacial time scale, has been controlled mainly by the details of the atmospheric circulation. Over the geologically short run the atmosphere has led and the oceans have followed.

Variations in solar output. Meteorologists agree that solar energy is the prime mover of the earth's atmosphere. The atmosphere receives most of its energy in the lower latitudes and dissipates it by means of various exchange mechanisms (including oceanic circulation) that effect a net transfer of "excess" heat to higher latitudes. The general circulation of the atmosphere is a response to and consequence of the gross distribution of solar energy received by the rotating earth, complicated considerably by differences in the nature, elevation, temperature, and roughness of its surface. In spite of these variations, the mean state of the atmosphere is such as to give rise to temperature gradients along the meridians and hence to latitudinal climatic belts, with a warm zone in the equatorial region and cold ones near the poles. Assuming a fixed rotational axis and taking into account the basic controlling factors, it is difficult to imagine a circumstance in which the overall distribution of temperature could be otherwise.

Clearly, changes in solar radiation might be expected to produce variations in atmospheric behavior, including swings toward conditions favoring glaciation. The question is whether such changes actually took place in the past. Unanimity is lacking, but many astronomers appear to regard the sun as essentially unchanging, at least in terms of the average total radiational output over long periods of time (Brandt and Maran 1972; Eddy 1977). On the other hand, that there are short-term fluctuations, as measured by sunspot and solar-flare activity, seems firmly established, and there is mounting observational evidence that the so-called solar constant (the amount of solar radiant energy striking a unit surface at the top of the atmosphere oriented perpendicular to the solar beam) undergoes small but measurable and possibly regular changes (Schneider and Mass 1975). The relationship between such changes and weather and climatic variations is far from clear, but that such a relationship exists seems probable (Wilcox 1976).

In the absence of an unambiguously documented connection between short- and/or long-term solar variability and atmospheric behavior tending toward glaciation, this causal factor is accorded an auxiliary position in our model. It does not work alone, but is an important component in a system of determinants. Variations of 1–2% in the solar constant, operating over a several-hundred-year cycle, are assumed to affect the earth's atmosphere, but no significant overall change in the total radiational output of the sun is presumed to have taken place over the course of much of earth history.

Volcanic activity. It has been known for many years that particulate matter could influence the behavior and temperature of the atmosphere (Humphreys 1964). Variations in atmospheric turbidity, including those induced by injection of volcanic "dust," have been held responsible for a number of observed short-term temperature fluctuations (Damon and Kunen 1976; Newell and Weare 1976), and speculations concerning long-term effects of sustained volcanic activity on the global heat balance are numerous. But as with most other facets of the glacial problem, definitive statements about the climatic significance of such activity are not available.

There is no unanimity regarding even the primary effect of a "volcanic dust veil." Some students of the problem postulate a decrease in surface temperatures produced by scattering and absorption of several percent of the direct solar beam; others suggest an increase, mainly because of an increase in downward infrared radiation at the top of the lower atmosphere (for general discussions see Lamb 1970; Pollack et al. 1976). The majority of contemporary investigators seem to have concluded that the net effect will be a cooling of the earth's surface, although the recent work of Idso (1973; Idso and Brazel 1977) introduces a salutary note of caution.

In the absence of clear-cut guidelines from the specialists, and in the presence of persuasive evidence of short-term influences (Hansen, Wang, and Lacis 1978), we assume here that the addition of large amounts of volcanic dust will result in an overall cooling of the lower atmosphere. Persistence of such cooling would have an effect on at least some phases of the hydrologic cycle and thus might influence the initiation and perpetuation of glaciation.

Difficulties quickly arise, however, if one attempts to find a simple cause-and-effect relationship between all known periods of prolonged volcanism and past glaciations. The geologic record from early Precambrian time to the present, as noted by Bryson (1974) and others, shows random volcanic activity. These sporadic periods of volcanism have but infrequently coincided in time with documented glaciations. There is some evidence suggesting an increased tempo of explosive volcanism over the past 2 million years (Kennett and Thunell 1975), although that interpretation has been challenged (Ninkovich and Donn 1976), but the volumetrically impressive Mesozoic and Tertiary volcanic activity in much of western North America (King 1969), for example, evidently had little effect in producing and sustaining major glaciations on this continent. Volcanic activity is therefore judged to have been inadequate, by itself, to initiate major continental glaciations. In conjunction with other factors, however, it may act as one critical element leading to the onset of glaciation.

Variations in surface albedo. Many students of the problem of glacial origins have looked for answers in surface-albedo changes, particularly those caused by fluctuations in the area of the earth blanketed by snow and ice (Barry, Andrews, and Mahaffy 1975). Systematic satellite imaging has made possible almost day-to-day mapping of the snow and ice cover, and the extent to which that cover changes over relatively short periods (1–3 years) has proved to be remarkable. Since, as emphasized by Kukla and Kukla (1974), the "location and duration of snow and pack-ice fields constitute the most important seasonal variable in the earth's heat balance," the possible significance of long-lasting albedo changes to climatic fluctuations is readily understandable.

In a model atmosphere, reducing the surface albedo by partially or completely eliminating the polar ice caps has produced calculated changes in equatorial and high-latitude temperatures of +2° to +17°C. Thus, as Sellers (1969) states, "albedo manipulation at either pole would have worldwide repercussions." The absolute values of such numbers are far less important than the fact that theoretical considerations and empirical evidence converge to suggest that potentially critical changes in atmospheric temperature can be generated by albedo variations.

The initial effect of a change in surface albedo on the atmospheric heat balance would probably be almost undetectable; only a sustained increase could play an influential role in large-scale climatic variability. On the other hand, observational evidence suggests that relatively rapid year-to-year weather changes may be traced to "sudden" increases or decreases in the snow- and ice-covered area of the earth. That such sudden variations can and do take place is strikingly illustrated by the 12% increase measured in 1971 by Kukla and Kukla (1974). These authors noted that "only seven similar occasions would be needed to establish the pleniglacial surface albedo."

Several investigators have proposed that the primary requirement for initiation of glaciation was a minor reduction in atmospheric temperatures, particularly summer temperatures (Mitchell 1965; Barry, Andrews, and Mahaffy 1975), and that climatic fluctuations of the past differed only in degree and not in kind from short-term variations of the present. If we assume this to be the case, what is wanted is a trigger, terrestrial or extraterrestrial, to induce a modest drop in warm-season temperatures. While there are those who would disagree, a sustained increase in surface albedo, producing a self-accelerating feedback mechanism leading to atmospheric cooling (Fairbridge 1972), could provide the necessary trigger. I believe that this has indeed been the case, that *only* an increase in surface albedo could have functioned on the required time scale to give rise to the observed effects.

Variations in the earth's orbit. James Croll suggested in 1875 that climatic change is caused by periodic variation in latitudinal and seasonal insolation—the amount of solar radiation received by a given area—produced not by variation in total solar output but by changes in the earth's orbit. The effects of changes in the earth's orbit have been much disputed ever since, but one theory that has much appeal for many students of recent earth history is the so-called radiation or Milankovitch theory (Milankovitch 1930; Vernekar 1972). Based upon accepted principles of celestial mechanics, it builds upon the fact that systematic changes in the orientation and inclination of the earth's axis, coupled with long-term changes

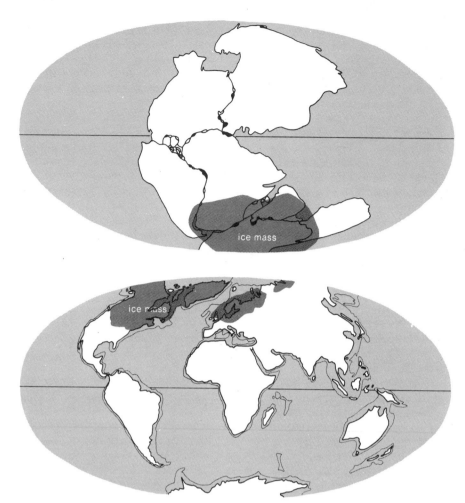

Figure 3. The Permo-Carboniferous glaciations covered the southern end of Pangaea when the supercontinent was centered at high southern latitudes (*above*). The late Cenozoic glaciation covered the northern parts of North America and Eurasia when those continents were high in the northern latitudes, in their present positions (*below*).

in the eccentricity of the earth's orbit, lead to variations in seasonal insolation at any given latitude on a cyclic, recurring basis. Differences of opinion regarding this factor have revolved largely around the magnitude of such changes and their possible influence on gross atmospheric behavior.

Most observers have regarded orbital changes as inadequate to *initiate* glaciation. If this were not the case, geological evidence of past ice ages presumably would be more widespread, both in time and place, whereas the available record strongly suggests geographical clustering of major continental glaciations scattered somewhat irregularly through time. It is reasonable to suppose, however, that once glaciation has been initiated (for whatever reason), orbital changes would favor alternation of glacial and interglacial conditions (Emiliani and Shackleton 1974). That orbital changes were responsible

for major fluctuations in the Pleistocene ice sheets has been persuasively argued by Hays, Imbrie, and Shackleton (1976), and the changing geometry of the earth's orbit is here considered to be the primary cause of glacial-interglacial variations once an ice age has been triggered.

Changes in the latitudes of continents. I believe that continental drift is the mechanism that has led to underlying conditions favorable for development of major glaciation, and that major continental glaciations have occurred only when large land masses have been in high latitudes. Fundamental to the argument is the proposition that a climatic *potential* for glaciation has always existed in the higher latitudes. What has been needed to get it started is the presence of continents in the right places (Fig. 3).

Alfred Wegener (1966) and a number

of other observers noted quite early the coincidence of the occurrence of major glaciations (late Cenozoic and earlier) with the presence of continents in high latitudes. Ironically, as has been pointed out by Hallam (1975) and Gould (1977), although much of the evidence now considered reasonable, rational, and "obvious" was just as valid and compelling 60 years ago as it is today, it has taken almost half a century for the intellectual seeds scattered by Wegener, Holmes, Taylor, du Toit, and a few others to find receptive scientific soil.

In the last decade or so, Tanner (1965), Hamilton (1968), Crowell and Frakes (1970), and Donn and Shaw (1977) have devised models of glaciation in which continental drift has been given as the primary or at least a prominent cause. And Fairbridge (1972) has written, "under processes related to plate tectonics, seafloor spreading and continental drift, from time to time during geological history, major land masses have come to coincide with the polar climatic belts, major seaways have become blocked and the Earth's liquid heat-control mechanism breaks down. . . . The result is an ice age."

Movement of North America and Eurasia into their present latitudinal positions can therefore be supposed to have been the basic, underlying factor leading to conditions favoring late Cenozoic glaciation. The paleomagnetic evidence persuasively indicates the general northward drift of these land masses since the beginning of the breakup of Pangaea some 200 million years ago (Dietz and Holden 1970; Oxburgh 1974; Donn and Shaw 1977), and there are enough hints in the incomplete geologic record to suggest the strong possibility of high-latitudinal continental locations during earlier glaciations. There is general agreement that the Permo-Carboniferous glaciations of Pangaea took place on a continent centered in high southern latitudes (Crowell and Frakes 1970; Gould 1977).

A synthetic model

The glacial model described here has a number of potential virtues, not the least of which is its relative simplicity. It violates none of the known or inferred physical laws governing atmospheric behavior, nor does it require sudden, gross climatic changes for which there is little meteorological justification. And its provisions are generally in accord with the evidence at hand.

The *primary* requirement for initiation of glaciation on planet earth, then, is believed to be the positioning of continents in high latitudes. Given this geographical situation, the factor held to trigger a glacial episode is an increase in the earth's surface albedo. What are the conditions giving rise to a comparatively sudden increase in surface albedo?

There is moderately reliable historical evidence that coincidence in time of an increase in explosive volcanic activity (yielding an intensified dust veil) with a decrease in total solar radiational output can generate a small but significant decline in atmospheric temperatures, which, in turn, would favor an increase in snowfall in a number of critical areas. Continuation of the chance juxtaposition of the effects of the two factors for a few decades, or hundreds of years, would produce a further drop in temperature and an increase in the earth's snow and ice cover, as a result of which the surface albedo would increase. A self-sustaining feedback mechanism would thus be established: temperatures in the lower atmosphere would continue to drop, more precipitation would occur as snow, and an episode of glaciation could have been triggered.

The "Little Ice Age" of the seventeenth through the nineteenth century provides an instructive though short-lived example of the climatic consequences of just such a combination of events. Sunspot activity declined precipitously from about 1645 to 1715 (see Schneider and Mass 1975; Eddy 1976), apparently decreasing total solar radiational output; frequent explosive volcanic activity occurred (Lamb 1970), with striking effects on the luminosity of the moon during total eclipses (Hédervári 1975); and a significant lowering of the snowline and increase in regional snow cover took place on upland plateaus in north-central Baffin Island, Canada (Barry, Andrews, and Mahaffy 1975). Temperatures fell worldwide, and the stage seemed set for initiation of another glacial event. Yet global atmospheric temperatures generally rose after about 1730, and the Little Ice Age was apparently terminated by a climatic amelioration some 70 years ago. Evidently the influence of one or more additional factors is needed to prolong a surface albedo increase for a period sufficient to sustain a major glaciation.

Vernekar's compilation of long-period variations of incoming solar radiation (1972) reveals a probable explanation. The geometry of the earth-sun system is such that over the past several thousand years, wintertime insolation in the upper latitudes of the northern hemisphere has been high—relative to conditions prevailing 8–18 thousand years ago (Hays, Imbrie, and Shackleton 1976). The general warming influence over thousands of years of high wintertime insolation would appear to be sufficient to overcome short-term cooling effects, over hundreds of years, produced by the chance combination of an increase in volcanic dust with a decrease in solar radiation. Little ice ages can develop under these circumstances, but big ones can't.

Renewal of continental glaciation by a sustained increase in surface albedo must await the return of the next period of low northern hemispheric wintertime insolation. This will occur in about 8–10 thousand years, and at that time northern hemispheric summertime insolation will also be low. The glacial episode that should then ensue may well be spectacular.

Five of the seven possible causal factors—continental latitude, solar variability, orbital-geometry changes, volcanic activity, and surface-albedo increase—can thus be intertwined to construct a glacial model. What of the other two? The role of mountain-building is considered to be only coincidental, not critical. Mountain-building and relative or absolute increases in continental elevations might well intensify the effects of ongoing glaciation, but they are not required, in this model and in others, to initiate it. Similarly, changes in the behavior of oceans and their physical and chemical characteristics, on the time scale involved (a few hundred to a few thousand years), are believed to be the result of glaciation already under way, not a cause of a renewed ice age. Significant change in oceanic circulation, brought about by drift of the continents into different locations, simply cannot occur rapidly enough to figure importantly as an immediate initiator of glaciation.

To recapitulate, movement of continental land masses into high latitudes (and thus into climatic belts in which glaciation may be initiated and sustained) creates the fundamental condition required for ice ages to develop. Random fluctuations in volcanic activity and possibly regular variations in solar radiation occasionally act to reinforce one another, thus bringing about an increase in surface albedo, a lowering of surface temperature, and the triggering of an incipient ice age. If a little ice age evolves during a low-insolation phase of the Milankovitch cycle (and the timing of this will vary, depending upon whether most of the land is in the high-latitude parts of the northern or southern hemisphere), a major glacial episode will be generated. Little ice ages that develop during periods of high wintertime insolation in the appropriate hemisphere are doomed to relatively rapid extinction. Given the intrinsic controlling factors, the earth appears at present to be locked into an ice-age condition, with orbital changes dictating repetition of the glacial-interglacial cycle so long as much of North America and Eurasia remain in comparatively high latitudes.

Implications

Certain environmental consequences should logically devolve from the model of glacial causes outlined above. And reasonable explanations, all springing naturally from the proposed interweaving of causes leading inexorably to glaciation, can be offered for a number of enigmatic paleoenvironmental indicators.

Granting a sustained surface albedo increase, several significant atmospheric changes should follow. Temperatures will drop, snow accumulations will reach critical size in parts of northeastern North America, eastern Siberia, highland Scandinavia, and in higher mountains around the world, and glaciers will develop and expand. Glacial expansion will lead to a further decline in atmospheric temperatures, caused both by the growing volume of ice and an associated increase in surface albedo. Outside of glaciated areas, lower mean temper-

atures and less precipitation would not be unexpected.

As more water is converted to ice, sea level will fall and the relative elevation of the continents will increase, producing a further drop in surface temperatures. Growing ice masses in the higher latitudes will affect the general atmospheric circulation, tending in the northern hemisphere to shift the belt of prevailing westerlies equatorward. As a result of increases in the volume of sea ice, changes in ocean-basin geometry brought about by falling sea level, and variations in atmospheric circulation, there are inevitable changes in oceanic characteristics and circulation that reinforce climatic tendencies already well established.

Reconstruction of the environment of the ice age 18 thousand years ago, based largely on study of the deep-sea core record (CLIMAP project members 1976), imply conditions not unlike those deducible from the model proposed here: cooler and drier unglaciated continental areas, southward displacement of the zonal westerlies in the northern hemisphere, and sea-surface temperatures some 2–4°C lower in the North Pacific and North Atlantic. A more recent study utilizing measurements of the deuterium/hydrogen ratio in ancient wood cellulose (Yapp and Epstein 1977) reached roughly comparable conclusions: lower ocean temperatures prevailed, and cooler summers were experienced over the ice-free parts of North America, at the time of the most recent glacial maximum.

While, at the right time and place, glaciation is a self-sustaining, snowballing process, ultimately climatic limits must be imposed as ice masses migrate into lower latitudes and altitudes, a point stressed some time ago by Tanner (1965). Deflection of storm tracks away from centers of accumulation by the growing continental glaciers will bring about reduction of moisture supply and a condition of "starvation." Without continuous volume addition from source regions, glacial margins stagnate and finally recede. Comparatively short-term advances and retreats ("stades" and "interstades" in contemporary jargon) will thus succeed each other during a period of major glaciation, and the location of

ice margins will constantly fluctuate.

Major glacial episodes come to an end, apparently very rapidly (Bryson et al. 1969; Prest 1969), when wintertime insolation in the appropriate hemisphere rises to a high and a critical change in the earth's heat balance is achieved. The overall pattern that emerges, one supported by considerable geological evidence, is alternation of relatively long glacial intervals (80–100 thousand years) with rather brief periods (10–15 thousand years) of interglacial respite.

The ultimate cause of glaciation is thus seen to be movement of continents into appropriate latitudes. Ice ages are not preceded by a sudden, worldwide drop in atmospheric temperatures, they *produce* such a drop. The normal global climate isn't necessarily warmer, but only nonglacial. And much of the fossil evidence upon which the time-honored concept of Tertiary "cooling" has been founded could be nothing more than a reflection of drifting of what are now the northern-hemisphere land masses and ocean floors toward the pole and hence into cooler climes.

Two recent interpretations of present and past distributions of certain climatically sensitive sediments (Drewry, Ramsay, and Smith 1974; Gordon 1975) point compellingly to the conclusion that overall patterns of atmospheric circulation, and in particular the latitudinal climatic zones, have remained essentially unchanged throughout the last 600 million years. The model of glacial causes described here fits comfortably into the global pattern envisioned by these studies.

References

Barry, R. G., J. T. Andrews, and M. A. Mahaffy. 1975. Continental ice sheets: Conditions for growth. *Science* 190:979–81.

Beckinsale, R. P. 1973. Climatic change: A critique of modern theories. In *Climate in review*, ed. G. McBoyle, pp. 132–51. Houghton Mifflin.

Brandt, J. C., and S. P. Maran. 1972. *New horizons in astronomy.* Freeman.

Brooks, C. E. P. 1970. *Climate through the ages: A study of the climatic factors and their variations.* Dover (unabridged reprint of 1946 rev. ed.).

Bryson, R. A. 1974. A perspective on climatic change. *Science* 184: 753–60.

Bryson, R. A., W. M. Wendland, J. D. Ives, and

J. T. Andrews. 1969. Radiocarbon isochrones on the disintegration of the Laurentide ice sheet. *Arctic and Alpine Res.* 1:1–14.

CLIMAP project members. 1976. The surface of the ice-age earth. *Science* 191:1131–37.

Cline, R. M., and J. D. Hays, eds. 1976. Investigation of late Quaternary paleoceanography and paleoclimatology. *Memoir Geol. Soc. Am.* 145.

Croll, J. 1875. *Climate and time in their geological relations: A theory of secular changes of the earth's climate.* London: Edward Stanford.

Crowell, J. C., and L. A. Frakes. 1970. Phanerozoic glaciation and the causes of ice ages. *Am. J. Sci.* 268:193–224.

Damon, P. E., and S. M. Kunen. 1976. Global cooling? *Science* 193:447–53.

Dietz, R. S., and J. C. Holden. 1970. Reconstruction of Pangaea: Breakup and dispersion of continents, Permian to present. *J. Geophys. Research* 75:4939–56.

Donn, W. L., and D. M. Shaw. 1977. Model of climate evolution based on continental drift and polar wandering. *Geol. Soc. Am. Bull.* 88:390–96.

Drewry, G. E., T. S. Ramsay, and A. G. Smith. 1974. Climatically controlled sediments, the geomagnetic field, and Trade Wind belts in Phanerozoic time. *J. Geology* 82:531–53.

Eddy, J. A. 1976. The Maunder minimum. *Science* 192:1189–1202.

———. 1977. A practical question in astronomy. Review of *Possible relationships between solar activity and meteorological phenomena,* ed. W. R. Bandeen and S. P. Maran. *Science* 195:670–71.

Eddy, J. A., P. A. Gilman, and D. E. Trotter. 1977. Anomolous solar rotation in the early 17th century. *Science* 198:824–29.

Emiliani, C., and J. Geiss. 1955. On glaciations and their causes. *Geologische Rundschau* 46:576–601.

Emiliani, C., and N. J. Shackleton. 1974. The Brunhes epoch: Isotopic paleotemperatures and geochronology. *Science* 183:511–14.

Fairbridge, R. W. 1972. Climatology of a glacial cycle. *Quat. Research* 2:283–302.

Flint, R. F. 1957. *Glacial and Pleistocene geology.* Wiley.

———. 1971. *Glacial and Quaternary geology.* Wiley.

Gordon, W. A. 1975. Distribution by latitude of Phanerozoic evaporite deposits. *J. Geology* 83:671–84.

Gould, S. J. 1977. The continental drift affair. *Natural History* 86:12–17.

Hallam, A. 1975. Alfred Wegener and the hypothesis of continental drift. *Sci. Am.* 232:88–97.

Hamilton, W. 1968. Cenozoic climatic change and its cause. In *Causes of climatic change,* ed. J. M. Mitchell, Jr., pp. 128–33. Am. Meteorol. Soc., Meteorological Monographs 8, no. 30.

Hansen, J. E., W.-C. Wang, and A. A. Lacis. 1978. Mount Agung eruption provides test of global climatic perturbation. *Science* 199:1065–68.

Hays, J. D., J. Imbrie, and N. J. Shackleton. 1976. Variations in the earth's orbit: Pacemaker of the ice ages. *Science* 194:1121–32.

Hédervári, P. 1975. Great volcanic eruptions and the luminosity of the moon during total eclipses. *Universitá di Ferrara Memorie Geopaleontologiche* 4:151–61.

Humphreys, W. J. 1964. *Physics of the air.* Dover (republication of 3rd ed., 1940).

Idso, S. B. 1973. Thermal radiation from a tropospheric dust suspension. *Nature* 241:448–49.

Idso, S. B., and A. J. Brazel. 1977. Planetary radiation balance as a function of atmospheric dust: Climatological consequences. *Science* 198:731–33.

Kennett, J. P., and R. C. Thunell. 1975. Global increase in Quaternary explosive volcanism. *Science* 187:497–503.

King, P. B. 1969. The tectonics of North America. Discussion to accompany the tectonic map of North America, scale 1:5,000,000. U.S.G.S. Professional Paper 628.

Kukla, G. J., and H. J. Kukla. 1974. Increased surface albedo in the northern hemisphere. *Science* 183:709–14.

Lamb, H. H. 1970. Volcanic dust in the atmosphere: With a chronology and assessment of its meteorological significance. *Phil. Trans. Royal Soc. Lond.* (A) 266:425–533.

———. 1972. *Climate: Present, past and future,* vol. 1: *Fundamentals and climate now.* London: Methuen.

Manabe, S., and T. B. Terpstra. 1974. The effects of mountains on the general circulation of the atmosphere as identified by numerical experiments. *J. Atmospheric Sci.* 31:3–42.

Milankovitch, M. 1930. Mathematische Klimalehre und astronomische Theorie der Klimaschwankungen I, part A. In *Handbuch der Klimatologie.* Berlin: Borntrager.

Mitchell, J. M., Jr. 1965. Theoretical paleoclimatology. In *The Quaternary of the United States,* ed. H. E. Wright and D. G. Frey, pp. 881–901. Princeton Univ. Press.

———. 1968. Concluding remarks. In *Causes of climatic change,* ed. J. M. Mitchell, Jr., pp. 151–59. Am. Meteorol. Soc., Meteorological Monographs 8, no. 30.

Namias, J. 1969. Seasonal interactions between the North Pacific Ocean and the atmosphere. *Monthly Weather Review* 97:173–92.

———. 1972. Large-scale and long-term fluctuations in some atmospheric and oceanic variables. In *The changing chemistry of the oceans,* ed. D. Dryssen and D. Jagner, pp. 27–48. Wiley.

Newell, R. E., and B. C. Weare. 1976. Factors governing tropospheric mean temperature. *Science* 194:1413–14.

Ninkovich, D., and W. L. Donn. 1976. Explosive Cenozoic volcanism and climate implications. *Science* 194:899–906.

Oxburgh, E. R. 1974. The plain man's guide to plate tectonics. *Proc. Geol. Assoc.* (London) 85:299–357.

Pollack, J. B., O. B. Toon, C. Sagan, A. Summers, B. Baldwin, and W. Van Camp. 1976. Volcanic explosions and climatic change: A theoretical assessment. *J. Geophys. Research* 81:1071–83.

Prest, V. K. 1969. Retreat of Wisconsin and Recent ice. *Map Geol. Surv. Can.* 1257A.

Sachs, H. M. 1976. Evidence for the role of the oceans in climatic change: Tests of Weyl's theory of ice ages. *J. Geophys. Research* 81:3141–50.

Savin, S. M. 1977. The history of the earth's surface temperature during the past 100 million years. In *Annual review of earth and planetary sciences,* vol. 5, ed. F. A. Donath, pp. 319–55. Annual Reviews, Inc.

Schneider, S. H., and C. Mass. 1975. Volcanic dust, sunspots, and temperature trends. *Science* 190:741–46.

Sellers, W. D. 1969. A global climatic model based on the energy balance of the earth-atmosphere system. *J. Applied Meteorology* 8:392–400.

Steiner, J., and E. Grillmair. 1973. Possible galactic causes for periodic and episodic glaciations. *Geol. Soc. Am. Bull.* 84:1003–18.

Tanner, W. F. 1965. Cause and development of an ice age. *J. Geology* 73:413–30.

Vernekar, A. D. 1972. Long-period global variations of incoming solar radiation. Am. Meteorol. Soc., Meteorological Monographs 12, no. 34.

Wegener, A. 1966. *The origin of continents and oceans.* Dover (translation and reprint of 4th rev. ed., 1929).

Weyl, P. K. 1968. The role of the oceans in climatic change: A theory of the ice ages. In *Causes of climatic change,* ed. J. M. Mitchell, Jr., pp. 37–62. Am. Meteorol. Soc., Meteorological Monographs 8, no. 30.

Wilcox, J. M. 1976. Solar structure and terrestrial weather. *Science* 192:745–48.

Yapp, D. J., and S. Epstein. 1977. Climatic implications of D/H ratios of meteoric water over North America (9,500–22,000 B.P.) as inferred from ancient wood cellulose C-H hydrogen. *Earth Planet. Sci. Letters* 34:333–50.

PART 2 *Climates of the Past*

Richard K. Bambach
Christopher R. Scotese
Alfred M. Ziegler

Before Pangea: The Geographies of the Paleozoic World

Pre-Pangean configurations of continents and oceans can be reconstructed according to our knowledge of geologic processes still operating in the modern world

Although the earth is known to be 4,600 million years old, most of our geologic information is from rocks formed in the last 570 million years, the time during which organisms have left a good fossil record. This span is subdivided into the traditional geologic time scale (Fig. 1). During this time, life colonized the land, the vast majority of deposits of fossil fuels accumulated, and all of the present mountain ranges of the earth formed. While this was going on, continental drift—a result of plate-tectonic processes—caused the geography of the earth to change constantly.

The paleogeographic history of the last 240 million years, the Mesozoic and Cenozoic eras, is well understood, because this part of the geologic record is relatively well preserved and the pattern of dated magnetic reversal stripes on the ocean floor allows us to reposition the continents as they were during this interval. Reconstructions of continental positions by

Richard K. Bambach is Associate Professor of Paleontology at Virginia Polytechnic Institute and State University. During 1978–79 he was Visiting Associate Professor at the University of Chicago. His research interests cover a wide range of topics in paleoecology. Christopher R. Scotese is a Research Assistant in the Department of Geophysical Sciences at the University of Chicago. He is a specialist in the use of computer graphics for depicting paleogeography and is currently studying the paleomagnetism of Paleozoic rocks. Alfred M. Ziegler is Professor of Geology at the University of Chicago. He has published in the areas of community paleoecology, stratigraphy, and paleogeography. All three authors have been working on an atlas of paleogeographic maps. Address: Richard K. Bambach, Department of Geological Sciences, Virginia Polytechnic Institute and State University, Blacksburg, Virginia 24061.

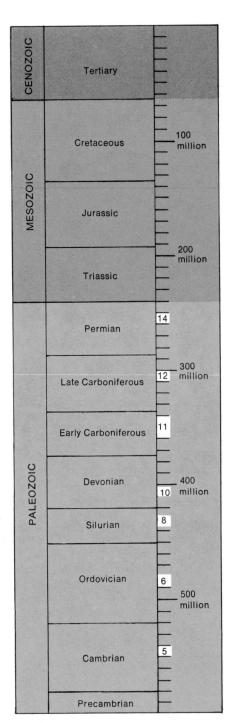

Figure 1. Geologic time. (White blocks represent time spanned by the indicated map.)

Dietz and Holden (1970a,b), Van der Voo and French (1974), and Smith and Briden (1977) show that about 240 million years ago the present continental blocks were grouped together into a "supercontinent" called Pangea. The most recent quarter billion years of geologic time has seen the breakup of Pangea, the formation of the "new" Atlantic and Indian Ocean basins, and the collision of some of the fragments of Pangea to form the Afro-Eurasian landmass, the nucleus of a new supercontinent. These events have been documented by data collected by projects such as the Deep Sea Drilling Project (Nierenberg 1978).

But because no pre-Pangean geographic relationships are preserved, the positions and paleogeography of the continents in the preceding 330 million years, the Paleozoic Era, is much more difficult to determine. The pioneering effort at mapping the positions of continental blocks in the Paleozoic was a set of four maps by Smith, Briden, and Drewry (1973), based solely on paleomagnetic information. The first reconstructions utilizing paleoclimatic and tectonic information combined with paleomagnetic data were presented in a study of the Silurian by Ziegler, Hansen, and others (1977) and another covering times from the start of the Paleozoic to the Cenozoic prepared in the Soviet Union (Zonenshayn and Gorodnitskiy 1977a,b). These maps identified most of the separate paleocontinents of the Paleozoic and proposed a logical, consistent sequence for Paleozoic plate motions.

The emerging pattern of geographic change during the Paleozoic allows us to return to a uniformitarian view of

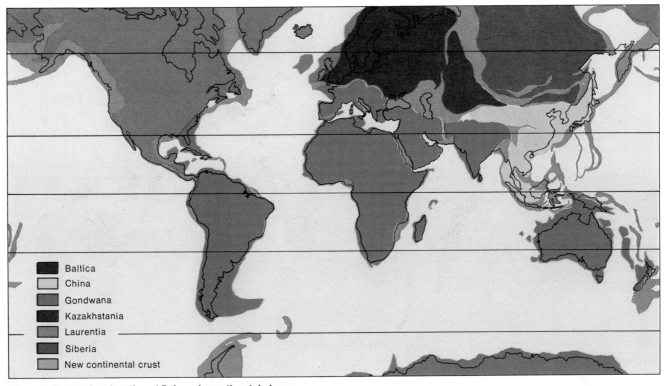

Figure 2. The modern location of Paleozoic continental pieces

Baltica
China
Gondwana
Kazakhstania
Laurentia
Siberia
New continental crust

earth history. No longer do we need to invoke improbable concepts of borderlands, land bridges, and worldwide shifts in climatic equilibrium to explain the seemingly odd locations in the modern world of ancient salt deposits, coral reefs, and evidence of glaciation or the widely scattered occurrence of similar fossil flora and fauna. Rather, we can understand how these features are the result of

movements of the continents themselves, by geologic processes still operating today.

Making paleogeographic reconstructions

Reconstructions of the changing Paleozoic world are made by identifying areas that acted as separate continents, positioning these paleo-

continents in their correct orientations, compiling data indicative of geographic and climatologic features, and interpreting the distribution of environmental conditions on each paleocontinent. The interpretation of Paleozoic geography requires synthesis of data derived from many fields of geology. Paleomagnetism has served as the cornerstone of this work (McElhinney 1973), but paleoclima-

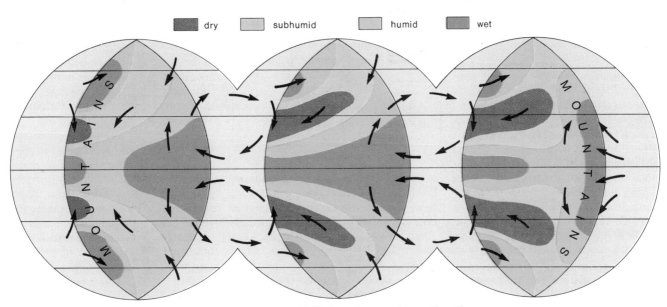

dry   subhumid   humid   wet

Figure 3. Idealized model of climatic conditions as a function of latitude and geographic configuration

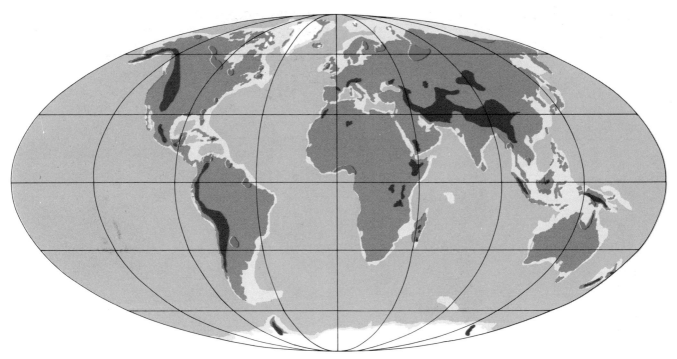

Figure 4. Generalized geography of the modern world

tic, paleobiogeographic, and tectonic data are also vital in compiling detailed reconstructions. Our methods and details of the results have been presented in a series of recent papers (Ziegler, Hansen, et al. 1977; Ziegler et al. 1977, 1979; Scotese et al. 1979), in which the basic data for the following reconstructions were presented.

Geography changes constantly. If too great a period is covered by a single reconstruction, the geographic conditions will be represented by too broad an average to give a clear picture of actual conditions at any particular time. The 12 geologic periods shown in Figure 1 are too long to be summarized by general reconstructions. These periods, however, are divisible into 75 stages, a stage corresponding to the smallest subdivision of time-equivalent rocks recognized worldwide. Each stage lasted from 5 to 15 million years, and since plates move at a rate of 2–8 cm/yr, changes in the relative and absolute positions of the continental blocks during such time spans are contained enough to be realistically summarized. A stage is thus a useful basis for a single reconstruction, though shoreline positions and the extent of glaciation shown are still averages of changing conditions within a stage.

The continental blocks of the Paleo-

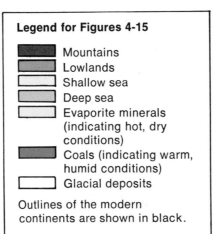

Legend for Figures 4-15

- Mountains
- Lowlands
- Shallow sea
- Deep sea
- Evaporite minerals (indicating hot, dry conditions)
- Coals (indicating warm, humid conditions)
- Glacial deposits

Outlines of the modern continents are shown in black.

zoic that collided to form Pangea were not the same as the continents that formed as Pangea split apart during the Mesozoic or as those which exist today. Identifying ancient continents requires the recognition of features that mark the outlines or margins of these paleocontinents. A major achievement of plate-tectonic theory has been the definition of criteria for recognizing these ancient continental boundaries (Mitchell and Reading 1969; Dewey and Bird 1970; Dickenson 1970; Burk and Drake 1974; Burke et al. 1977). The torn and rifted margins of once-associated continents are represented by certain geologic features, especially belts of basaltic

igneous rocks associated with elongate basins bordered by normal faults. Other features, especially belts of andesitic igneous rock and mountain belts with strongly folded rocks, represent the deformed margins of continental blocks under which oceanic crust was subducted as lithospheric plates moved together. Where such belts cross a continent, such as the Ural Mountains in Eurasia, two former continents appear to have collided and been sutured together.

We recognize six major paleocontinents during various parts of the Paleozoic: Gondwana, Laurentia, Baltica, Siberia, Kazakhstania, and China. Figure 2 shows the parts of the modern continents that belonged to each of these paleocontinents and areas of the present continental crust that have accreted to the margins of the continents during Paleozoic, Mesozoic, and Cenozoic times.

Gondwana was a supercontinent composed of what are now South America, Florida, Africa, Antarctica, Australia, India, Tibet, Iran, Saudi Arabia, Turkey, and southern Europe. Laurentia comprised most of modern North America and Greenland, with the addition of Scotland and the Chukotski Peninsula of the eastern USSR. The missing parts of eastern North America were either part of Gondwana or associated with

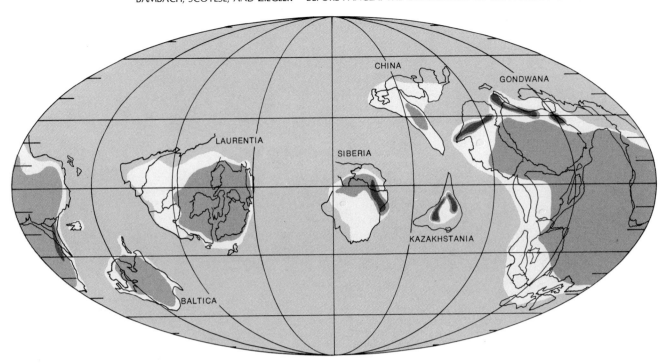

Figure 5. Late Cambrian (550–540 million years ago)

small microcontinents similar to modern New Zealand. Baltica was composed of Russia west of the Ural Mountains, Scandinavia, Poland, and northern Germany. Siberia from the Urals east to the Verkhoyanski Mountains was a separate continent in the Paleozoic. Its southern margin crossed Asia north of present Kazakhstan and south of Mongolia. Kazakhstania was a triangular continent centered on what is now Kazakhstan, with one part extending up between the Urals and southwestern Siberia and another part extending east between the Altai on the north and the Tien Shan Mountains on the south. China is a complex area that may have been subdivided into more than one block at times in the Paleozoic, but there are general similarities which imply that the pieces were not widely separated. For now, we treat all of southeast Asia, including China, Indochina, and part of Thailand and the Malay Peninsula, as a single continent.

Positioning the continents is at the heart of any reconstruction of world paleogeography. Assigning correct positions to paleocontinents involves both the latitude-longitude location and the orientation of the continents relative to the appropriate paleonorth direction. We believe, for instance, that in the Silurian period Siberia was centered at about lat. 30° north and

was rotated 180° from its modern orientation, so that its present Arctic coast faced south.

Paleomagnetic information is the basis for all continental positioning. Paleomagnetism provides direct quantitative evidence for both latitude and the north-south orientation of the paleocontinents. Although the magnetic poles may stray from the rotational poles, the earth's magnetic field has a north-south polarity and maintains an alignment that over geologic time on the average parallels the rotational axis. The lines of magnetic force also vary in their inclination to the earth's surface as a function of latitude, being vertical at the magnetic poles and parallel to the surface at the equator. The remnant magnetism of adequately preserved rock samples indicates the azimuth direction to the north (or south) magnetic pole and the latitude of the rock at the time the magnetism was imposed. Thus, from oriented samples we can determine the latitude and orientation of continental blocks for times in the past. Paleoclimatic indicators serve as independent checks on latitude and as guides to latitude when paleomagnetic information is not available.

Longitudes are determined by integrating biogeographic relationships and plate-motion constraints. Al-

though it is not possible to assign absolute longitude (relative to the prime meridian) in the Paleozoic, the whole interval is bracketed by times when relative longitude can be determined within narrow limits. At the time of our earliest reconstruction, in the Late Cambrian, all the continents except Baltica and China straddled the equator (see Fig. 5) and, with their surrounding oceans, occupied much of the space available in the equatorial belt. Since their relative order in sequence around the globe is clear from biogeographic evidence, space constraints alone fix their longitudinal positions within rather narrow limits.

At the end of the Paleozoic (Fig. 14), most of the continental blocks were grouped together in the supercontinent of Pangea. The spatial fit of the continents at this time tells us their relative longitudes quite precisely. Thus, with the endpoints of the Paleozoic well defined, we can follow a regular pattern of plate motion as the plates, bearing the separate continents which ringed the equator in the Cambrian, shifted and brought the continental blocks together to form the pole-to-pole mass of Pangea by the Permian.

The final task in preparing paleogeographic reconstructions is interpreting the geographic features and

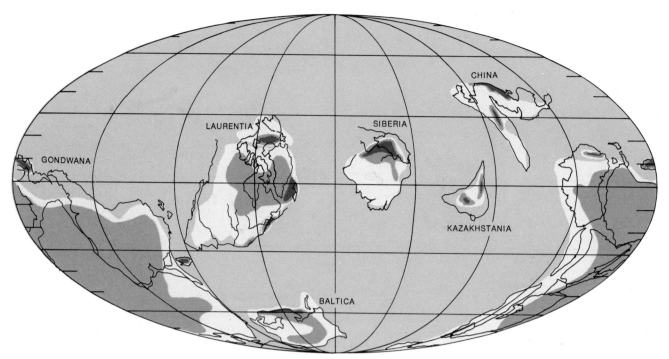

Figure 6. Middle Ordovician (490–475 million years ago)

distribution of environmental conditions on the continental blocks. In our reconstructions we identify regions that were highlands and mountains, lowlands and floodplains, coastal areas, shallow and deep marine platforms, and submarine-slope and deep-sea environments. Environmental conditions are interpreted from the processes known to have formed particular types of rocks.

Sands and coarser-grained detrital sediments are deposited in high-energy environments such as stream channels and along beaches, while fine-grained sediments such as clays and muds accumulate in low-energy environments such as floodplains, lagoons, and deep offshore marine environments. Limestones typify warm, shallow marine conditions at some distance from a source of land-

derived detrital sediments. Sands and muds, for example, are present along both the Atlantic and Gulf coasts of modern North America where rivers bring them to the shore, but in the Florida Keys and the Bahamas—areas far removed from sources of detrital sediments—limestones are accumulating from the skeletal secretions of abundant marine life. Fossils help differentiate terrestrial,

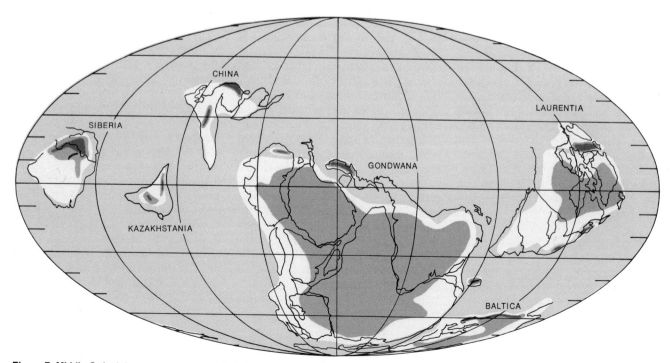

Figure 7. Middle Ordovician, view of earth rotated 180° from the view in Figure 6

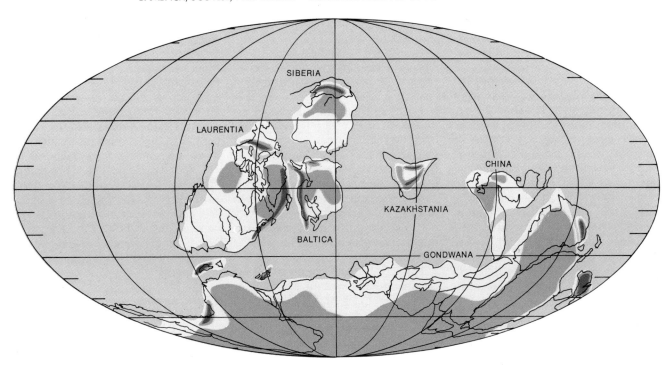

Figure 8. Middle Silurian (435–430 million years ago)

freshwater, and marine deposits, as do suites of sedimentary structures unique to particular depositional settings.

Some sediments reflect climatic conditions. Detrital sediments are indicative of humid environments. Coals formed in freshwater swamps where temperature and rainfall were both adequate for abundant plant growth. Hot, dry shorelines and protected basins along coasts in desert regions are the sites for deposition of salts and other evaporite minerals from brines left by evaporating seawater or playa lakes. In cold environments, glaciers leave distinctive deposits, called *tillites,* with features imposed by ice flow and the peculiar conditions of freezing and thawing at the melting edges of the ice. In areas where nutrients from deep water are supplied in abundance to surface waters, such as zones of coastal upwelling, the sediments may include cherts, phosphorites, or deposits rich in organic material, the result of high biologic productivity.

Geographic features can also be inferred from the type and structure of rock bodies. Delta complexes and

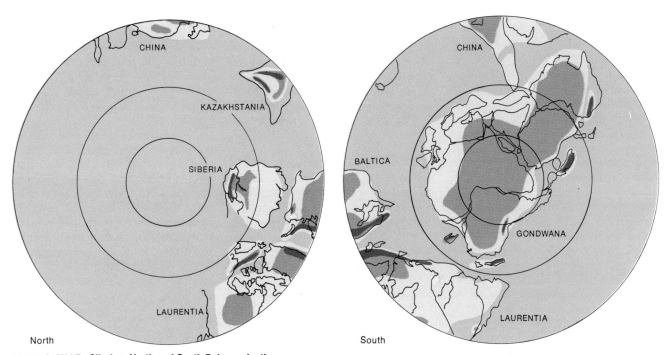

Figure 9. Middle Silurian, North and South Polar projections

deep-sea fans have typical internal features and distinctive three-dimensional forms. Belts of folded and thrust-faulted strata associated with metamorphic rocks and igneous intrusions represent areas that were mountain ranges at the time the deformation and metamorphism occurred, even if they are eroded to lowlands today. Andesitic igneous rocks were erupted along continental margins where subduction of oceanic crust occurred, as is happening in western South America and in the Aleutians today. Basalts were erupted and normal fault basins developed where rifting, associated with seafloor spreading, started, as in the modern African rift valleys or in Iceland.

The distribution of environments on the modern earth is related to the systematic arrangement of topographic features associated with plate-tectonic processes and to the climatic belts produced by atmospheric and oceanic circulation. These patterns are not random now, and they weren't in the past. The circulation of the atmosphere and oceans is an unchanging geophysical system produced by heat-transfer activities on a rotating earth as it is warmed by the sun. Despite the changes in geography over geologic time, some of which have produced local anomalies, the general climatic pattern of the earth is fixed over time. The present earth serves very well as the basis for a general model of global climate structure.

The modern world as a model

The sun is a stable main-sequence star and probably has not altered its intensity of radiation very much in the last billion years. The earth has always been spherical and has orbited at the same distance from the sun. Therefore the heat budget of the earth, with excess heating in the equatorial region, heat transfer by atmospheric and oceanic circulation toward the poles, and cooling in the polar regions, can be regarded as fixed. Although the earth's rotation has been slowed somewhat by tidal friction, it has not decreased by more than about 15% in the last half billion years. This means that the Coriolis effect, which deflects motions to the right in the Northern Hemisphere and to the left in the Southern

Hemisphere, has remained nearly constant.

The influence of the Coriolis effect on the airflow generated by heating and cooling of the earth's surface creates a latitudinally zoned pattern of atmospheric circulation (Strahler and Strahler 1978). The equatorial belt is characterized by hot, humid conditions with irregular surface winds, because the heated air is primarily expanding and rising rather than moving in a particular direction. Dry air, cooling and contracting in the upper atmosphere, sinks toward the surface of the earth at about lats. 30° north and south of the equator (the horse latitudes), causing dry climates with irregular surface winds. The sinking air flows out from this belt both toward the equator and toward the poles. The surface flow toward the equator is deflected by the Coriolis effect into strong prevailing easterly winds (the trade winds). The surface flow of air poleward is also deflected by the Coriolis effect and becomes the prevailing westerly winds of the temperate belts at lats. 40°–50° north and south of the equator.

In the polar regions cooling causes the air to contract and sink. This cold polar air flows as surface winds toward the equator, and these winds are deflected by the Coriolis effect into easterly winds in high latitudes. The polar front, where the equatorially trending polar air intersects the poleward flow of temperate air, is a region of high precipitation. The 23.5% tilt of the earth's rotational axis generates the seasonal fluctuations in climate as the earth orbits the sun. These fluctuations are most pronounced in the high temperate latitudes, where the polar front shifts back and forth, causing seasonal temperature changes from below freezing to above.

The interaction of this regular pattern of atmospheric circulation with the land-sea distribution (geography) of the earth produces a climatic regime that is also regular but does not consist of simple latitudinal belts. Except for locations very near the equator or at extremely high latitudes, most continents have quite different climates on opposite coasts at the same latitude (Fig. 3). The tropical humid zone widens toward the eastern sides of continents, where moisture is brought from the ocean by

prevailing easterly winds, but it is narrow on the western sides, confined to the narrow zone of intense heating where surface air is rising and losing moisture. The arid belts rise in latitude across the continents from west to east. They extend closer to the equator in the belts of easterly winds on the rain-starved western sides of continents and extend poleward in the midlatitude regions of prevailing westerly winds on their similarly rain-starved eastern sides. As with the wet eastern sides of continents in the tropics, the wet belts in temperate latitudes are much broader on the windward, west-facing margins of continents.

In the modern world these features are seen in the extensive arid belt which rises from low latitudes of the Sahara in eastern Africa to the high-latitude Gobi desert of western Asia. The humid regions of southeast Asia (Indochina, Burma, south China) are at the same latitude as the Sahara in the trade-wind belt (of easterlies), while France and Austria are at the same latitude as the Gobi Desert in the belt of prevailing westerlies. In the Southern Hemisphere, the Atacama Desert extends equatorward on the west coast of South America into latitudes occupied by the Amazon rain forest to the east, and the south Chile rainy zone is at the same latitude as the dry Argentinian pampas. The Andes create intense contrasts even across the narrow South American continent because of the rain shadow caused by their height.

The distribution of features in the modern world serves as a model for understanding climate patterns in the past. Because so many features of sedimentary rocks reflect climatic influence, we can map major climatic features on paleogeographic reconstructions. And the fact that climatic systems are predictable from their modern distribution means, as we have said, that we can use paleoclimatic data to cross-check paleomagnetic determinations of latitude. Paleoclimatic features also provide latitudinal information about regions and times for which paleomagnetic data are not available.

The Paleozoic world

Figures 5, 6, 7, 8, 10, 11, 12, and 14 are reconstructions of world paleogeography at seven times during the

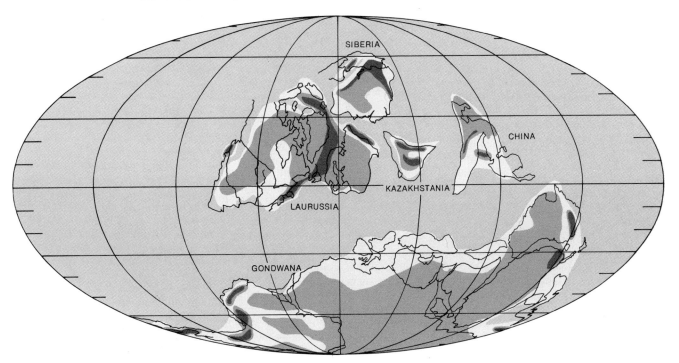

Figure 10. Late Early Devonian (410–405 million years ago)

Paleozoic. These reconstructions show the evolution of the Paleozoic world from the Late Cambrian through the Late Permian. We have used a Mollweide projection, which is an equal-area projection that avoids the excessive area expansion in higher latitudes of the familiar Mercator projection. Although there is angular distortion near the edge of the map at high latitudes, the Mollweide projection also shows the entire earth's surface, including the polar regions, which a Mercator projection cannot do.

The pageant of geographic change through the Paleozoic is profound. None of the Paleozoic geographies is similar to that of our modern earth. In the modern world (Fig. 4) the continents are grouped into three extensive north-south masses—the Americas, Europe/Africa, and eastern Asia through Indonesia to Australia— which partly isolate three equatorially centered oceans—the Pacific, Atlantic, and Indian oceans. The modern continents are mostly emergent, with only narrow continental shelves flooded by shallow seas. There are close connections between almost all the continents surrounding the North Pole, and the small Arctic Ocean is virtually landlocked by the belt of high-latitude land extending from Greenland across northern North America and Siberia to Scandinavia.

Antarctica covers the South Pole. Extensive high mountain belts are characteristic, most prominently the Alps-Caucasus-Himalayas system, extending across the Eurasian supercontinent, and the Andes–Rocky Mountains system, extending from Tierra del Fuego to Alaska in the Americas. Areas forming climatically sensitive sediments today are shown on the reconstruction for correlation with modern climate distribution and for comparison with the distribution of ancient deposits.

The ancient Late Cambrian world (Fig. 5) contrasts sharply with the world of today. In the Late Cambrian the continents of the Paleozoic were isolated from each other and dispersed around the globe in low tropical latitudes. The ocean basins were extensively interconnected and the polar regions were occupied by broad open oceans. There was no land above lats. 60° north or south. Shallow seas had transgressed onto the low-lying continental platforms earlier in the Cambrian period and covered large areas of Laurentia, Baltica, Siberia, Kazakhstania, and China in the middle Late Cambrian. The major highlands were in northeastern Gondwana (Australia and Antarctica today), eastern Siberia, and central Kazakhstania. Erosion had reduced the topography of Laurentia and Baltica to low levels.

During the Ordovician (Fig. 6) and Silurian (Fig. 8) Gondwana moved southward from its Late Cambrian position on the equator halfway around the globe from Siberia to a position straddling the South Pole. This change can be followed by comparing the latitude of Gondwana in Figures 5, 7, and 9. The oldest record of glaciation in the Paleozoic era is of Late Ordovician age in what is now the Sahara Desert: tillites were deposited at the time that this part of Gondwana was actually crossing the South Pole.

During the movement of the plate containing Gondwana, Baltica also shifted position along a parallel path from the Cambrian to the Silurian. The two paleocontinents may have been located on one lithospheric plate. Siberia also shifted from equatorial to north temperate latitudes during the early Paleozoic.

These movements were marked by mountain building along the eastern margin of Laurentia (the Taconic orogeny) in the middle and late Ordovician and later mountain building in western Baltica during the Silurian (the Caledonian orogeny), as the ocean between Laurentia and Baltica closed. Shallow seas were widespread on the continents throughout most of the Ordovician and Silurian. Climatic zonation is detectable in the distri-

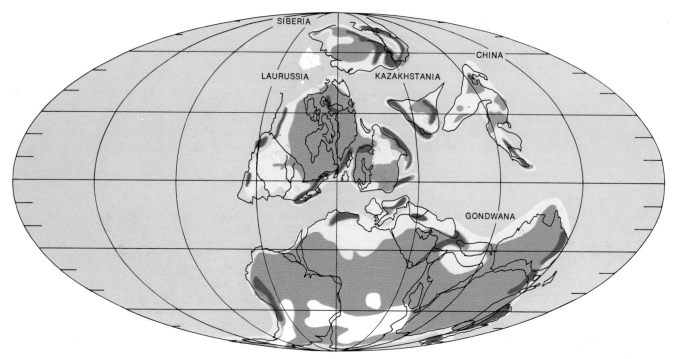

Figure 11. Middle Early Carboniferous (360–340 million years ago)

bution of evaporite deposits, which are concentrated between lats. 15° and 30° north and south, as in the modern world.

Frigid polar climates are not detectable in the Cambrian or most of the Ordovician. This may be because the open polar oceans of these times were always warmed by oceanic circulation, or it may simply be that the lack of polar land prevented preservation of a geologic record of polar climates. As mentioned above, the first land areas to enter polar latitudes, the North African portion of Gondwana, was glaciated in the Late Ordovician. Marine faunas of relatively low diversity—and therefore probably from temperate climates—are known from the higher south latitude areas of Gondwana in the Ordovician and Silurian, and a distinctive low-diversity fauna is found in the Silurian along the margin of Siberia, which first moved as far north as lat. 40° north.

Major reorganization of world geography is apparent by the Silurian. The shift southward of Gondwana from its Cambrian position had opened the former North Polar ocean basin until it not only circled the world at high northern latitudes but extended in an unbroken expanse southward across the equator to high southern latitudes. The former South Polar ocean basin of the Late Cam-

brian had been displaced to middle southern latitudes by the shifts of Gondwana, Baltica, and Siberia and had become a partly enclosed basin between Baltica, Kazakhstania, and Gondwana. This was the start of the development of the Tethys Sea, a region characterized by distinctive marine faunas throughout the Late Paleozoic and Mesozoic.

By the Early Devonian (Fig. 10), Laurentia and Baltica had collided to form a larger continent, Laurussia. The collision began in the Late Silurian with the Caledonian orogeny in northwestern Baltica. Mountain-building continued into the Devonian as the Acadian orogeny in eastern Laurentia. These uplands along the suture between the formerly separate paleocontinents were located in the equatorial belt, and large volumes of detrital sediments were eroded from them. Nonmarine fluvial sediments covered large parts of eastern North America (the "Catskill Delta") and northern Europe; these sediments were deeply weathered and are stained red from iron oxides—hence the name Old Red Sandstone in Great Britain. Although land plants had begun to evolve in the Silurian, it is in these tropical nonmarine Devonian deposits that abundant larger fossil land plants first appear.

The Early and Late Carboniferous

reconstructions (Figs. 11 and 12) show that Gondwana continued to move across the South Pole and entered the same hemisphere as Laurussia, closing the ocean between them. Their collision in the Late Carboniferous resulted in the Hercynian orogeny, which extended across central Europe, and the Alleghenian orogeny in eastern North America. Baltica had begun colliding with Laurentia in the Silurian and Devonian at a location relative to Laurentia far to the south of the position it occupied later. Repositioning took place in the Carboniferous, during the collision between Laurussia and Gondwana, when what had been Baltica was displaced northward along a series of faults extending from coastal New England, across Newfoundland, and through Scotland along the zone of weakness at the original suture.

From the Silurian through the Carboniferous, Siberia moved to high latitudes and Kazakhstania and China moved westward. By the Late Carboniferous, Kazakhstania and Siberia had collided and all the paleocontinents were clustered tightly as Pangea began to take shape (Fig. 13). Mountain belts extended along the suture between Laurussia and Gondwana (the southern Appalachian and Hercynian belts) and along the reactivated suture where

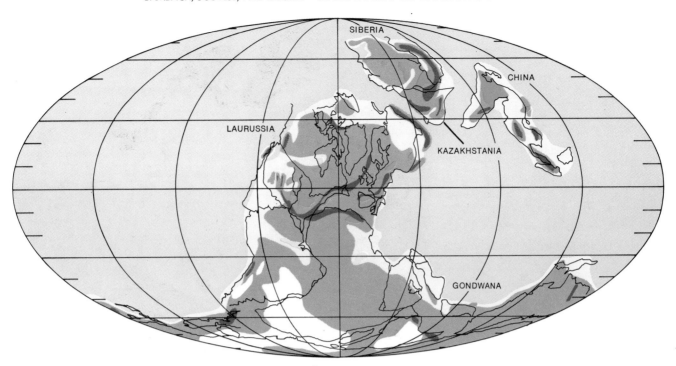

Figure 12. Middle Late Carboniferous (310–300 million years ago)

Baltica sheared northward relative to Laurentia (the northern Appalachian-Caledonide belt). Mountain systems marking subduction zones also developed during the Carboniferous on the eastern side of Baltica, as the ocean between Baltica and Kazakhstania-Siberia was closing, and in the Mongolian portion of Siberia, as the ocean between Siberia and China also closed.

The distribution of climatically indicative deposits in the Late Carboniferous (Fig. 12) reconstruction is particularly interesting. The great coal reserves of eastern North America, western Europe, and the Donetz Basin of the USSR lie in the equatorial zone. The coal swamps developed on marshy delta platforms built by rivers bringing detrital sediments from the adjacent mountain ranges.

The large volume of both detrital sediments and plant remains testifies to the high rainfall in this tropical belt. Plant fossils in the coals do not show strong seasonal growth rings, which implies that they grew under constantly warm rather than temperate seasonal conditions. The belt of tropical coals is narrow in the west and broadens to the east, as predicted by the climate models (see Fig. 3).

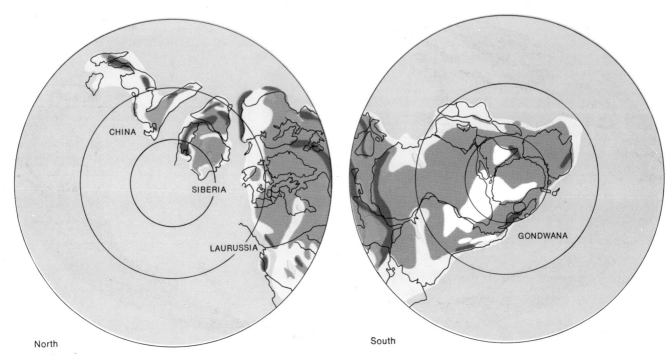

Figure 13. Middle Late Carboniferous, North and South Polar projections

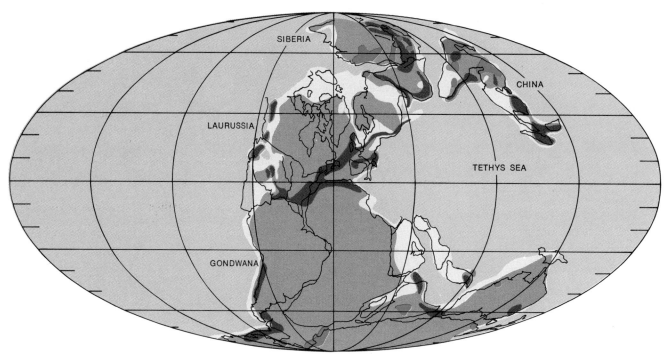

Figure 14. Early Late Permian (260–250 million years ago)

Evaporite deposits, indicative of low rainfall, occur in far western Laurussia at low latitudes and farther east between lats. 15° and 30° north and in a belt in Gondwana between lats. 20° and 30° south, again as expected from the climatic model. In the north temperate belt (lats. 40°–60° north) extensive detrital sediments and coals in Siberia and China indicate abundant rainfall. The seasonal nature of the climate of this belt is indicated by well-developed growth rings in plant fossils from Siberia. Seasonal growth rings, imposed by interruption of growth during cold winter months, are also found in plant fossils of both Carboniferous and Permian age from the south temperate latitudes of Gondwana.

Glaciation is recorded by widespread tillites in southern Gondwana. Ice sheets were present above lat. 60° south from the Early Carboniferous into the Early Permian. At their most extensive they flowed equatorward as far as the middle temperate latitudes, just as the Pleistocene glaciers flowed south from the Arctic regions of North America and Europe less than 100,000 years ago. The fact that the Carboniferous glacial deposits are

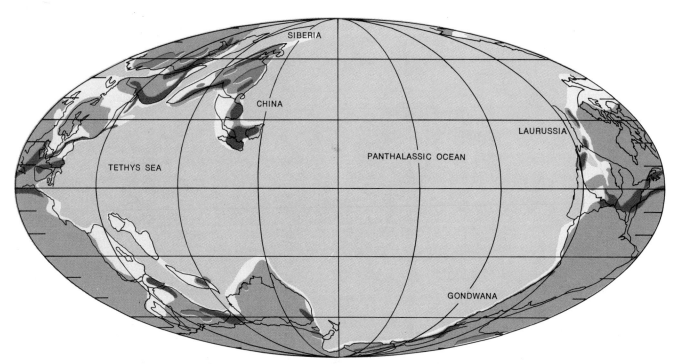

Figure 15. Early Late Permian, view of earth rotated 180° from the view in Figure 14

now found on continents scattered across half the earth's surface (from South America through southern Africa to Antarctica and India) was one of the strongest intimations of continental drift before the theory of plate tectonics.

Pangea was nearly assembled by the Late Permian (Fig. 14). It is worth noting, however, that a Pangea containing all continental blocks probably never formed completely. In eastern Gondwana during the Permian, rifting had begun that pulled Tibet, Iran, and Turkey away from the Tethyan margin of Gondwana as a separate, isolated block or blocks before China collided with the Mongolian region of the former Siberian block in the Triassic.

The effects of continental collision reach far beyond the simple suturing together of two formerly separate continental blocks. Collision deforms rocks through the entire thickness of the lithosphere along the colliding margins of the plates. The resulting folding and thrust faulting cause the rocks to "pile up" on themselves, thus thickening the continental crust in the zone of collision. Because lighter continental crust is buoyed up on the denser mantle, these belts of thickened, deformed continental crust form uplifted mountain belts.

And because the same mass of crustal material as formerly existed in undeformed, thinner crust is packed into these thickened belts, the area covered by continental crust is decreased during collision, just as the area covered by a rug is less if it is crumpled up against a wall rather than spread flat on a floor. The decrease in continental crustal area during collision is taken up by an increase in the area of the ocean basins, with a concomitant increase in their volume. This can cause a general lowering of sea level in relation to the continents, as shallow seas flooding continental platforms drain into the enlarging ocean basins.

The series of continental collisions through the late Paleozoic (Devonian–Permian) formed extensive mountain belts and decreased the total area of continental crust. As a result of this process and possibly also because of changes in the volume of mid-ocean ridges at the spreading centers, sea level was effectively lowered and most of the shallow seas that had flooded the low continental platforms through much of the Paleozoic were drained. The larger areas of exposed land contributed to increased climatic severity.

Permian terrestrial sediments indicative of arid and semiarid conditions were widespread. The mountains in the tightly sutured region between Laurussia and Gondwana were high enough to disrupt the subtropical easterly winds and create an intense rain shadow even in tropical latitudes, much as the Andes do in western South America today. Evaporites of Permian age extend to what was the equator in western Laurussia. Desert conditions also extended north of lat. 40° in eastern Laurussia on the side of the continent downwind from the westerly winds, just as deserts extend northward in eastern Asia today.

The clustering of the continents into Pangea had an extraordinary geographic consequence. An enormous single interconnected ocean developed (Fig. 15). This "world ocean," sometimes called Panthalassa, not only spanned the globe from pole to pole but extended for 300° of longitude at the equator, twice the distance from the Philippine Islands to South America across the modern Pacific. Circulation in this giant ocean had to have a major impact on Permian climates. For example, the equatorial currents driven by the trade winds flowed uninterrupted around five-sixths of the circumference of the earth. The east-facing (Tethyan) coast of Pangea, against which these currents impinged, must have been extremely warm. Ancient "Gulf Streams" would have circulated these warm waters into higher latitudes, causing an especially strong climatic asymmetry between eastern and western Pangea. The Permian was indeed a time of geographic extremes.

Paleogeography and geologic history

The old dictum "the present is the key to the past" has a rather specific meaning for geologic history. The processes of geology, including plate tectonics, have probably operated much as we observe them today over long spans of geologic time, and this is certainly the case if we accept reasonable variation in their rates. But although geographic features have always reflected the operation of processes and systems we observe today, the order of geographic change is a unique historical sequence. Thus the present *is* the key to understanding past processes, but it *is not* the key to describing past configurations.

Because plate-tectonic processes operate constantly, there has been no stable or even average geography of the earth during the past half billion years. Paleogeography changed continuously, passing from one extreme configuration through a series of intermediates to a different extreme, as illustrated by these maps, and then changing still further. The geographies of the Cambrian, Permian, and the present are simply single steps in this dynamic pattern.

The Cambrian, with its isolated, equatorially distributed continents and two polar but interconnected oceans, was totally unlike the Permian, with its single concentration of continents stretching from pole to pole and its immense, equatorially centered Panthalassic Ocean. The widespread shallow seas that flooded continental platforms in the Early Paleozoic contrast with the large proportion of exposed land in the Late Paleozoic. These two extremely different periods also differed totally from the modern world, which is the product of the breakup of Pangea into nearly interconnected north–south continental belts and large, semi-isolated ocean basins.

The modern world, of course, is just another transient stage in geologic history. Sound—though broadly outlined—predictions on the future course of geographic change indicate that the Atlantic Ocean, which has been opening for the past 150 million years as the Americas have been moving westward, will grow larger. The Atlantic coast of North America will probably develop into an Andean type of mountain system in the not-too-distant geologic future. There is very little subduction in the Atlantic today, but it is likely to begin in the next 50 million years, especially in the relatively old western North Atlantic.

On the other side of our continent, Baja California and the part of Southern California west of the San

Andreas Fault will continue on their present course of northwestward motion as part of the Pacific plate. California will not "fall into the sea," but the part west of the San Andreas fault may become a New Zealand-like small continental island moving away from mainland North America in the next 50 to 100 million years. Across the Pacific Ocean, Australia, which has been moving northward away from Antarctica for the last 40 million years, will move north past Indochina and may collide with China, Japan, or far eastern Russia. The world's largest ocean then will be the interconnected Indian-Antarctic-South Pacific.

The historical development of our earth has followed a nonrepetitive path through time. The world of the Permian was a world alien to the one in which we live; the world of the Cambrian was equally alien, to the Permian as well as the present; and the world 100 million years from now will be alien to our present one. It is within this framework of changing geographies, with markedly different extreme configurations and long intervening transitions, that we must cast our ideas of geologic history.

References

Burk, C. A., and C. L. Drake, eds. 1974. *The Geology of Continental Margins*. New York: Springer-Verlag.

Burke, K., J. F. Dewey, and W. S. F. Kidd. 1977. World distribution of sutures: The sites of former oceans. *Tectonophysics* 40: 69–99.

Dewey, J. F., and J. M. Bird. 1970. Mountain belts and the new global tectonics. *J. Geophys. Res.* 75:2625–47.

Dickenson, W. R. 1970. Relation of andesites, granites, and derivative sandstones to arc-trench tectonics. *Revs. of Geophysics* 8: 813–60.

Dietz, R. S., and J. C. Holden. 1970a. The breakup of Pangea. *Sci. Am.* 223(4):30–41.

———. 1970b. Reconstruction of Pangea: Breakup and dispersion of continents, Permian to present. *J. Geophys. Res.* 75: 4939–56.

McElhinny, M. W. 1973. *Paleomagnetism and Plate Tectonics*. Cambridge Univ. Press.

Mitchell, A. H., and H. T. Reading. 1969. Continental margins, geosynclines, and ocean floor spreading. *J. Geol.* 77:629–46.

Nierenberg, W. A. 1978. The deep sea drilling project after ten years. *Am. Sci.* 66:20–29.

Scotese, C. R., R. K. Bambach, C. Barton, R. Van der Voo, and A. H. Ziegler. 1979. Paleozoic base maps. *J. Geol.* 87:217–68.

Smith, A. G., and J. C. Briden. 1977. *Mesozoic and Cenozoic Paleocontinental Maps*. Cambridge Univ. Press.

Smith, A. G., J. C. Briden, and G. E. Drewry.

1973. Phanerozoic world maps. In *Organisms and Continents through Time*, ed N. F. Hughes, pp. 1–42. Paleontological Association Special Papers in Paleontology, no. 12.

Strahler, A. N., and A. H. Strahler. 1978. *Modern Physical Geography*. Wiley.

Van der Voo, R., and R. B. French. 1974. Apparent polar wandering for the Atlantic-bordering continents: Late Carboniferous to Eocene. *Earth Science Reviews* 10:99–119.

Ziegler, A. M., K. S. Hansen, M. E. Johnson, M. A. Kelly, C. R. Scotese, and R. Van der Voo. 1977. Silurian continental distributions, paleogeography, climatology, and biogeography. *Tectonophysics* 40:13–51.

Ziegler, A. M., C. R. Scotese, W. S. McKerrow, M. E. Johnson, and R. K. Bambach. 1977. Paleozoic biogeography of continents bordering the Iapetus (pre-Caledonian) and Rheic (pre-Hercynian) oceans. In *Paleontology and Plate Tectonics*, ed. R. M. West, pp. 1–22. Milwaukee Public Museum, Special Publications in Biology and Geology, no. 2.

———. 1979. Paleozoic paleogeography. *Ann. Revs. Earth and Planet. Sci.* 7:473–502.

Zonenshayn, L. P., and A. M. Gorodnitskiy. 1977a. Paleozoic and Mesozoic reconstructions of the continents and oceans, article 1: Early and Middle Paleozoic reconstructions. *Geotectonics* 11:83–94.

———. 1977b. Paleozoic and Mesozoic reconstructions of the continents and oceans, article 2: Late Paleozoic and Mesozoic reconstructions. *Geotectonics* 11:159–72.

Jack A. Wolfe

A Paleobotanical Interpretation of Tertiary Climates in the Northern Hemisphere

Data from fossil plants make it possible to reconstruct Tertiary climatic changes, which may be correlated with changes in the inclination of the earth's rotational axis

Anyone who has even a slight acquaintance with paleoclimatic literature is well aware that the last 1 to 1.5 million years of the Quaternary have been characterized by major episodic glaciations of the continents of the Northern Hemisphere, and hence the period is atypical of much of geologic time. A commonly accepted thesis on climates preceding Quaternary glaciation is that, from some time in the Late Cretaceous or early Tertiary (some 40–80 m.y. ago), when the earth's climate was characterized by generally higher temperatures and higher equability of temperature than now, both overall temperature and equability have gradually decreased, culminating in Quaternary glaciation. Further, some researchers have maintained that even as long ago as the early Tertiary, temperatures were only moderately higher than now, even at high latitudes. (See Table 1 for the geologic time span dealt with in this article.)

An increasing accumulation of data from a multitude of sources has,

In 1957, Erling Dorf delivered the Ermine Cowles Case Memorial Lecture at the University of Michigan, a lecture sponsored by Sigma Xi. His lecture, on Tertiary climates from a paleobotanist's viewpoint, which was later published in American Scientist, *shows some parallels in conclusions with the present article, but much of the basic information Dorf accepted has undergone major revision by subsequent paleobotanical and stratigraphic work. A version of the present paper was also delivered as a Case Lecture in October 1975. Jack A. Wolfe was educated at Harvard and Berkeley and is a geologist at the U.S. Geological Survey. His interests are in Cenozoic floras of western North America and systematics and phylogeny of angiosperms. Address: Paleontology and Stratigraphy Branch, U.S. Geological Survey, 345 Middlefield Road, Menlo Park, CA 94025.*

however, largely negated such once commonly accepted theses. A significant warm episode during the Miocene (see Fig. 1) was first documented in Europe by Mai (1964) and has subsequently been substantiated in other regions such as Japan (Tanai and Huzioka 1967), western North America (Wolfe and Hopkins 1967; Addicott 1969), and New Zealand (Devereux 1967). Alpine glaciation is known to have begun in Alaska during the Miocene (Denton and Armstrong 1969; Plafker and Addicott 1976), when at least part of the Antarctic ice sheet was also present (Kennett 1977).

The most dramatic climatic event, however, occurred during the middle of the Tertiary. MacGinitie (1953) recognized that, if certain floras in Oregon were as close in time as some stratigraphic evidence indicated, a rapid and major climatic change must have occurred, a decrease in temperature that was considered significant but gradual by Nemjč (1964) and Zhilin (1966). Utilizing newly available radiometric ages, Wolfe and Hopkins (1967) demonstrated that this major climatic deterioration had occurred within 1 or 2 m.y.

This temperature decrease has subsequently been recognized in many regions. I had previously (1971) termed it the "Oligocene deterioration," but since recent work in relating the marine and nonmarine chronologies indicates that, in the widely accepted chronology based on marine plankton, the event occurred at the end of the Eocene, I will refer to it as the "terminal Eocene event." In the Southern Hemisphere, the terminal Eocene event is closely associated with the initiation of cold bottom water in the oceans (Kennett 1977),

while on the continents of the Northern Hemisphere the event is emphasized by a major decrease in equability of temperature (Wolfe 1971).

Foliar physiognomy

A thorough review of all pertinent paleoclimatic data for the Tertiary would be a lengthy and prodigious task. In this paper I will largely limit the discussion to the paleoclimatic data based on fossil plants from middle to high latitudes (<30°) of the Northern Hemisphere. For the Paleocene and Eocene, the North American data, which are based on leaf remains, are the most relevant. In Europe, the major Eocene floral sequence is based on fruit and seed floras. In eastern Asia, some Oligocene floras have been described (Tanai 1970), but most of the Paleocene and Eocene assemblages remain undescribed and unanalyzed (Tanai 1967).

There are several advantages in basing paleoclimatic interpretations on

Table 1. The subdivisions of the Cenozoic Era

		Million years ago
Quarternary		
Holocene		.012
Pleistocene		1.5
Tertiary		
Pliocene	} Neogene	5
Miocene		23
Oligocene	} Paleogene	33
Eocene		53
Paleocene		65

Figure 1. From the estimated percentages of species with entire-margined leaves in four locations in North America, it is inferred that a sharp drop in mean annual temperature took place in the early Oligocene and has continued—at least at high latitudes—to the present day. At middle latitudes mean annual temperature has, overall, not changed since the Oligocene. Dotted intervals indicate that leaf-margin data are either lacking or not considered reliable.

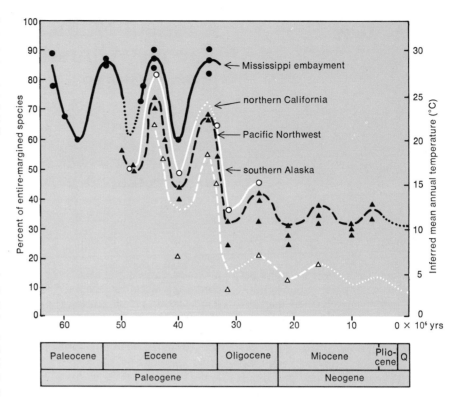

the physical aspects (or physiognomy) of fossil leaf assemblages. The physical characteristics of vegetation occupying similar climates in widely separated regions are highly similar, although the regions may have only a few taxa in common. Among the most conspicuous of such characteristics are the similarities in the appearance of foliage. On the other hand, vegetation occupying dissimilar climates in one region typically has different physical characteristics, although many taxa may occur throughout the region.

Thus the environment tends to select plants that have certain physical aspects for a given climatic type, whether this climatic type is separated by oceans or, presumably, by major periods of time. Just as we can expect the Tropical Rain forest in Africa to have the physical characteristics of Tropical Rain forest in other regions of the world, so we can also expect the present Tropical Rain forest to have the physical characteristics of the Tropical Rain forest of the Eocene.

On the other hand, the Tropical Rain forest of Indonesia has a floristic composition markedly different from the composition of the Tropical Rain forest of Brazil. Such differences have resulted from a variety of historical factors, both geographic and evolutionary. These historical factors will also result in floristic differences between the Tropical Rain forest of the Eocene in a given region and any part of the modern Tropical Rain forest. Considering that the floristic composition of any vegetational type is continually undergoing change, then the determination of the vegetational type (and hence climatic type) represented by a fossil assemblage is best

accomplished by analyzing the physical characteristics of the assemblage rather than its floristic composition. And, the further back in time, the more dissimilar the floristic associations are to present associations and the more problematic become climatic inferences.

Among the most useful physiognomic characters of broad-leaved foliage are: type of margin, size, texture, type of apex, and type of base and petiole (see Fig. 2). In areas of high mean annual temperature and precipitation, for example, leaves typically have "entire" margins (i.e. lacking lobes or teeth), are large, are coriaceous ("leatherlike"—an indication of an evergreen habit), and have a high proportion of attenuated apices (i.e. "drip-tips," particularly common on lower-story plants); and a moderate number have cordate (heart-shaped) bases associated with palmate venation and joints ("pulvini") in the petiole—a combination of characters typically associated with the vine, or liana, habit.

The general correlation between type of leaf margin and climate was first

Figure 2. Physical characteristics of leaves largely represent adaptations to the environment and are thus good indicators of climate. The small (microphyllous) leaf (*top*), an alder leaf from southern Alaska, has an incised margin and is characteristic of cool climates. The vine leaf (*bottom*) from the Philippines has a swollen and jointed petiole, palmate venation, and a cordate (heart-shaped) base, as do the leaves of most vines. The attenuated tip (drip-tip) indicates a humid habitat, and large (mesophyllous) size and entire (smooth) margin are characteristic of most tropical plants. White bars represent 30 mm.

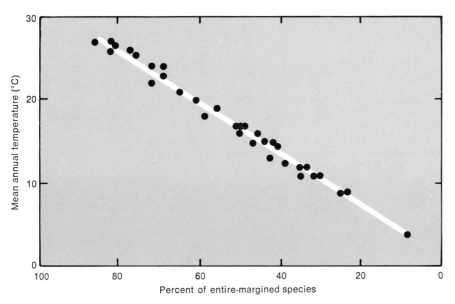

Figure 3. The percentage of species with entire-margined leaves in the humid to mesic broad-leaved forests of eastern Asia increases in direct proportion to the mean annual temperature of the particular forest.

documented by Bailey and Sinnott (1915) and has since been sporadically applied to interpretations of fossil assemblages. A recent compilation of analyses of woody vegetation in the humid to mesic (moderately humid) forests of eastern Asia has shown a strong correlation between the percentage of species with entire leaf margins and mean annual temperature (Fig. 3). Compilations of leaf-margin data of secondary vegetation—vegetation on disturbed sites that has not reached a climax stage—and of the broad-leaved element in coniferous forests do not display such a correlation.

Although leaf size is an important criterion in studies of extant vegetation, the application of this parameter to fossil assemblages is highly problematic, because leaves of different sizes may be differentially selected in the process of transport and preservation (Spicer 1975). Further, leaf-size changes can be related to precipitation and soils as well as to temperature. In the following discussion, I have used a generalized and modified version of the Raunkiaer (1934) system of leaf sizes: *mesophyll* for the larger mesophyll and larger classes, *notophyll* for the smaller mesophyll class (Webb 1959), and *microphyll* for the smaller classes.

The significance of physiognomy to the paleobotanist attempting paleoclimatic reconstructions is that the major physiognomic subdivisions of vegetation (which are partly based on

foliar characters) have been found to correspond closely with certain major temperature parameters (Wolfe, in press). Figure 4 shows that mean annual temperature (an approximation of heat accumulation) is of major significance in determining what type of vegetation prevails, as are warm-month means. Only two cold-month means are of major significance. The 1°C mean separates dominantly broad-leaved evergreen (above 1°C) from broad-leaved deciduous (below 1°C) forests; in the areas that have cold-month means between 1°C and −2°C, notophyllous broad-leaved evergreens occur as an understory element, and in regions of even greater winter cold, notophyllous broad-leaved evergreens are lacking. The 18°C cold-month mean—a commonly accepted boundary between "tropical" and "subtropical"—has no relevance to the distribution of vegetation.

Estimates of mean annual temperature can be based on the percentage of entire-margined species in a given fossil assemblage. More difficult to infer is the mean annual *range* of temperature, which, in some cases, can be estimated only within broad parameters. In other cases, however, mean annual range of temperature can be accurately inferred. For example, if two succeeding assemblages have the same leaf-margin percentage of 50% (mean annual temperature ~17°C), and if the younger assemblage is dominantly microphyllous and the older assemblage is domi-

nantly notophyllous, then reference to the framework of Figure 4 indicates that a mean annual range of temperature of 6°C was reached some time between the two assemblages. A second example is that of the Miocene Seldovia Point flora, which represents vegetation slightly inland from the coast of southern Alaska. Foliar physiognomic (as well as floristic) criteria indicate a mean annual temperature of 6–7°C (Wolfe and Tanai, in press). Other paleobotanical data indicate that the broad-leaved deciduous forest represented by the Seldovia Point flora merged with coniferous forest toward the coast. Again, reference to Figure 4 indicates a mean annual range of temperature of about 26–27°C.

Two major problems that have hampered many climatic inferences from paleobotanical data have been the lack of floras in even moderately close stratigraphic successions and the total misinterpretations of the age and climatic significance of high-latitude Tertiary floras. These misinterpretations arose from acceptance of the undocumented concept of an "Arcto-Tertiary Geoflora"—that the Eocene vegetation in Alaska represented temperate broad-leaved deciduous forest that, unchanged, gradually migrated southward to middle latitudes. In North America, both problems have, to a high degree, been overcome. In Alaska there are stratigraphic sequences of floras—many independently dated—that represent most of the Tertiary. In the Pacific Northwest, numerous floras—again, many in stratigraphic succession and/or independently dated—occur in early Eocene and younger rocks. In the Mississippi embayment region, an almost complete sequence of floras represents most of Paleocene and Eocene time. Almost all these floras represent coastal plain vegetation, and thus one major variable in interpreting the significance of paleoclimatic inferences—altitude—is held approximately constant. Certain floras from interior areas add other dimensions to paleoclimatic models, but the altitu-

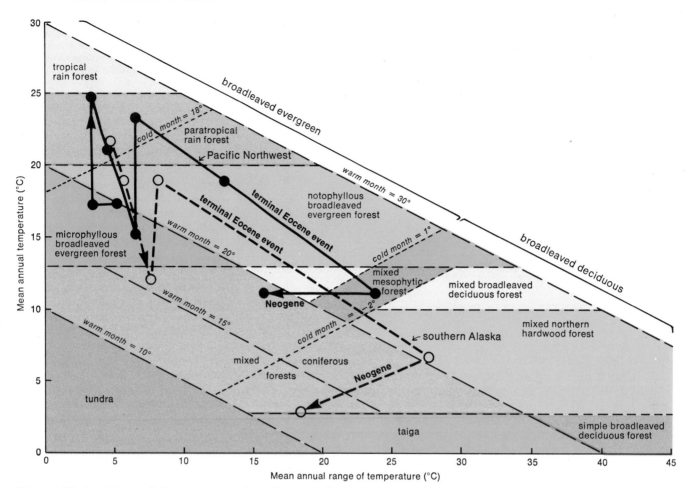

Figure 4. The humid to mesic forests of the Northern Hemisphere can be approximately circumscribed by various major temperature parameters. By comparing leaf assemblages in southern Alaska and the Pacific Northwest to the modern vegetation, we can infer the mean annual temperature and mean annual range of temperature for the assemblages. Major changes in temperature parameters are indicated for the time span between the middle Eocene and the Quaternary—showing a dramatic increase in mean annual temperature and an increase in mean annual range of temperature during the terminal Eocene event.

dinal factor introduces a problematic variable.

Paleocene and Eocene climates

The most complete sequence of Paleogene leaf floras in a small area is that of the Puget Group in western Washington (Wolfe 1968). The Puget assemblages extend from an estimated 50 m.y. ago (late early Eocene) up to about 34 m.y. ago (latest Eocene). In this sequence, the floras all contain numerous leaf species that have drip-tips and/or probable liana leaf physiognomy, and coriaceous (i.e. ≅ evergreen) texture dominates. Thus, all the assemblages represent vegetation that apparently grew under abundant year-round precipitation and would be classed as broad-leaved evergreen rain forests. Major changes in margin and size of the leaf assemblages occurred, however, during deposition of the Puget Group (Fig. 5). The leaf-margin data alone indicate major (perhaps 7–9°C) fluctuations in mean annual temperature, and, in a general manner, the leaf-size data also indicate climatic fluctuations. Sequences of floras in western Oregon, eastern Oregon, and northeastern California parallel the Puget sequence (Wolfe 1971).

The Puget leaf-size data are, however, possibly significant in a context other than mean annual temperature. In the lower part of the sequence, although the leaf-margin data do not significantly change, there is a pronounced movement from a notophyllous to a microphyllous forest. If mean annual temperature was approximately constant during this interval, then mean annual range of temperature must have decreased (i.e. equability of temperature increased). Indeed, the combination of physiognomic data indicates a mean annual range of temperature about half that at present in coastal western Washington, which is highly equable now in comparison to most other mid-latitude areas of the Northern Hemisphere. Although other workers have on questionable floristic interpretations (e.g. Berry 1914, p. 66–67) suggested that the Eocene was characterized by high equability, the Puget physiognomic data provide strong evidence that the Eocene was in fact highly equable.

One of the major corollaries of high equability during the Eocene has been generally overlooked, particularly by the proponents of the concept of an "Arcto-Tertiary Geoflora," who long argued that the Eocene vegeta-

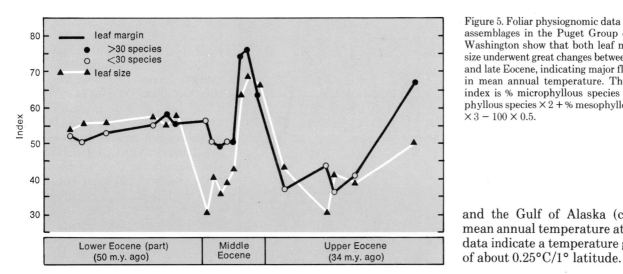

Figure 5. Foliar physiognomic data for Eocene assemblages in the Puget Group of western Washington show that both leaf margin and size underwent great changes between the early and late Eocene, indicating major fluctuations in mean annual temperature. The leaf size index is % microphyllous species + % notophyllous species × 2 + % mesophyllous species × 3 − 100 × 0.5.

and the Gulf of Alaska (ca. 22°C mean annual temperature at 60° N.) data indicate a temperature gradient of about 0.25°C/1° latitude.

The Alaskan Paleocene assemblages are exceedingly difficult to interpret climatically. In southeastern Alaska on Kupreanof Island (lat 57°) there are large assemblages that in all aspects of physiognomy represent Notophyllous Broad-leaved Evergreen forest (mean annual temperature approximately 18°C). Yet, only 2–4° latitude northward, the bulk of the broad-leaved evergreen element is unrepresented in the even larger assemblages of the Chickaloon and West Foreland formations. In features of foliar physiognomy such as margin and size, the Chickaloon assemblages would appear to correspond to Notophyllous Broad-leaved Evergreen forest (Wolfe 1972), yet the Chickaloon assemblages are dominantly broad-leaved deciduous, and even the broad-leaved deciduous element is not as diverse as in present temperate broad-leaved deciduous forests. The minor broad-leaved evergreen element includes palms and certain dicotyledonous families that are not to be expected in temperate broad-leaved deciduous forests. Thus, although almost all physiognomic characters and limited floristic data point to temperatures that should support dominantly broad-leaved evergreen vegetation, the vegetation was dominantly deciduous. Such assemblages occur throughout much of Alaska and Siberia north of latitude 60° during the Paleocene.

In the southeastern United States a large number of Paleogene leaf assemblages that represent coastal plain vegetation occur. Detailed work by many stratigraphers allows an accurate placement of these assemblages in stratigraphic sequence (Fig. 6). The oldest assemblages—early Paleocene (Midway Group)—are

tion in regions such as Alaska represented temperate broad-leaved deciduous forest. As in the now highly equable areas of the Southern Hemisphere or on tropical mountains, temperate and mesic broad-leaved deciduous forests could *not* have existed in the Northern Hemisphere during the Eocene. That is, the latitudinal temperature gradient would fall far to the left side of Figure 4—far from temperatures that would support temperate broad-leaved deciduous forest.

Notable also in the Puget analyses is that, during the late middle Eocene (ca. 45 m.y.), the vegetation was marginally Tropical Rain forest. A contemporaneous assemblage in northern California—the Susanville flora (lat 40°)—has a leaf-margin percentage of 82, concomitant with a leaf-size index of 79. The other physiognomic data are consistent with inferring the Susanville flora to be Tropical Rain forest. This indicates that Tropical Rain forest (and the 25°C isotherm) occurred at least 20° and possibly 30° poleward of the present northern limit.

Many assemblages that those who argued for an "Arcto-Tertiary Geoflora" interpreted as temperate broad-leaved deciduous forest are indeed that type; however, these assemblages are typically of Neogene age (cf. the radiometric data of Triplehorn et al. 1977), rather than Eocene, as they had supposed. In fact, only two small Eocene assemblages were known from Alaska until the last decade.

Collections from the Eocene at 60–61° latitude in the Gulf of Alaska region (Wolfe 1977) represent the latter half of the epoch. The late middle Eocene floras—correlative with the Susanville flora—represent Paratropical Rain forest and indicate the warmest climate (ca. 22°C mean annual temperature) of the Tertiary in Alaska (Wolfe 1972). The warmth indicated by the foliar physiognomic data is fully substantiated by the floristic evidence: included are feather and fan palms, mangroves, and members of other families now dominantly or entirely tropical (Wolfe 1977).

Recent geologic data (e.g. Jones et al. 1977) indicate that parts of southern Alaska were once at low latitudes and have drifted northward. The drift and accretion of these plate fragments to Alaska were, however, accomplished by the beginning of the Tertiary. The various major models of plate tectonics are unanimous in suggesting that, in general, western North America rotated southward during the Tertiary. That is, the paleolatitudes of these western North American floras were probably higher than the present latitudes of the fossil localities.

The latitudinal temperature gradient along the Pacific Coast of North America is today very moderate—about 0.5°C/1° latitude. During the late middle Eocene, however, the gradient was even lower. The Susanville (ca. 27°C mean annual temperature at 40° N.), the Puget (ca. 25°C mean annual temperature at 48° N.), those from Naborton and Mansfield, Louisiana. The physiognomy of these assemblages is clearly indicative of Tropical Rain forest (mean annual temperature about 27°C). The succeeding late Paleocene (lower part of Wilcox Group) assemblages represent Paratropical Rain forest—an indicated cooling consonant with data from the continental interior (Wolfe and Hopkins 1967). In the earliest Eocene assemblages (upper part of Wilcox Group), however, leaf size is reduced, the probable liana type of leaf is not as common as earlier, and drip-tips are uncommon; at the same time, the leaf-margin data suggest a warm interval.

How much of a hiatus exists between the Wilcox and Claiborne assemblages is uncertain, but I am assuming that most of the late early and early middle Eocene is missing, at least in the floral sequence. The large assemblages from Puryear, Tennessee, and Granada, Mississippi (the bulk of the "Wilcox flora" of various authors, but actually Claiborne in age; cf. Dilcher 1973a), are of late middle Eocene age. Dilcher (1973b) considered the climatic inferences based on such assemblages to be puzzling; however, the scarcity of probable lianas, the scarcity of drip-tips, and the small leaf size concomitant with a high leaf-margin percentage are characteristic of dry tropical vegetation (cf. Rzedowski and McVaugh 1966).

It is perhaps also significant that these Claiborne assemblages contain a diversity of Leguminosae, a family common in dry tropical vegetation today. Apparently a cooling occurred near the Claiborne-Jackson boundary, but the one cool assemblage (interestingly, once interpreted by Berry, 1916, to be of Pleistocene age) is unfortunately small. In any case, the Mississippi embayment sequence indicates a pronounced drying trend from the Paleocene into at least the middle Eocene.

In the continental interior, the Paleocene floral sequence also shows a definite cooling from the early into the late part of the epoch (Wolfe and Hopkins 1967; Wolfe, in press). Hickey (1977) suggests a renewed warming trend near the Paleocene-Eocene boundary, which would parallel the Mississippi embayment trend. As in that area, the interior

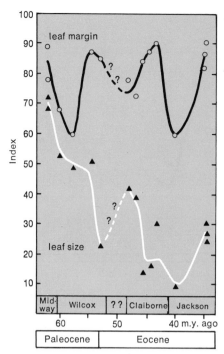

Figure 6. Indexes from foliar physiognomic data for Paleocene and Eocene assemblages in the Mississippi embayment region indicate a pronounced drying trend from the Paleocene into at least the middle Eocene.

Paleocene assemblages all indicate humid to mesic vegetation.

In the Eocene, however, the pattern in the interior becomes greatly complicated. The late early and early middle Eocene assemblages (the earliest Eocene assemblages are unstudied) from central and northern Wyoming represent definite mesic conditions, but the assemblages from southern Wyoming and adjacent Colorado and Utah represent pronounced dry conditions (MacGinitie 1969, 1974). How this situation is related to the presence of mountains and consequent rain shadows is uncertain. Today, the predominant sources of moisture for this region are southerly; one would expect the more southern area (southern Wyoming) to be moister if the Eocene circulation pattern were similar. Later Eocene leaf assemblages from this region are poorly known, except for the latest Eocene Florissant flora of central Colorado (MacGinitie 1953). The climatic significance of this flora is problematic because the altitude at which the Florissant beds were deposited is unknown.

To the west, a number of floras are known from an ancient uplifted area

that stretched from Nevada north into British Columbia. The known assemblages represent mesic coniferous forest. Two—the Princeton, British Columbia (Arnold 1955), and the Republic, Washington (Berry 1929)—are of early middle Eocene age and represent the same cool interval as documented in western Washington. The Copper Basin and Bull Run floras from northern Nevada (Axelrod 1966) are correlative with the late Eocene cool interval.

The Paleocene and Eocene floras from North America thus provide the basis for a number of climatic inferences. (1) An overall gradual warming took place from the Paleocene into the middle Eocene, with gradual cooling until the terminal Eocene event. (2) Cool intervals occurred during the late Paleocene, the late early to early middle Eocene, and the early late Eocene. The difference between the intervening warm intervals was, in mean annual temperature, about 7°C. (3) The cool intervals were about 4 to 5°C (mean annual temperature) warmer than the present. (4) Mean annual range of temperature during the middle Eocene was about half that of the present. (5) Mean annual range of temperature decreased from the early into the middle Eocene and possibly increased slightly until the end of the Eocene. (6) The latitudinal temperature gradient during the middle Eocene along the west coast of North America was about half that of the present. (7) The west coast of North America received abundant precipitation during that period. (8) The southeastern United States experienced a pronounced drying trend from the Paleocene into at least the middle Eocene.

Oligocene and Neogene climates

The most profound climatic event of the Tertiary took place at the end of the Eocene. In middle to high latitudes of the Northern Hemisphere, the vegetation changed drastically. Within a geologically short period of time, areas that had been occupied by broad-leaved evergreen forest became occupied by temperate broad-leaved deciduous forest. A major decline in mean annual temperature occurred—about 12–13°C at latitude 60° in Alaska and about 10–11°C at latitude 45° in the Pacific Northwest. Just as profound, however, was the shift in temperature equability: in the Pacific Northwest, for example, mean annual range of temperature, which had been at least as low as 3–5°C in the middle Eocene, must have been at least 21°C and probably as high as 25°C in the Oligocene (Fig. 4; Wolfe 1971).

One of the major aspects of early Oligocene floras at middle to high latitudes is their lack of diversity, which was followed by enrichment during the remainder of the Oligocene (Wolfe 1972, 1977). The lack of diversity would be expected following a major and rapid climatic change such as the one that characterized the terminal Eocene event—that is, few lineages were preadapted or could rapidly adapt to the new temperature extremes.

Although the late early to early middle Miocene warming has been recognized throughout the world, some evidence indicates a warm interval during the late Oligocene (see references cited by Wolfe 1971) and perhaps, to a lesser extent, a warming during the latest Miocene (Wolfe 1969; Barron 1973). These warm intervals, however, were not as warm in comparison to adjoining cool intervals as were the Paleocene-Eocene warm intervals.

The climatic trends following the terminal Eocene event, aside from the minor fluctuations, are of great significance. One trend that can be demonstrated in areas north of latitude 30° is an increase in equability, a trend that runs counter to putative models of Neogene climatic change (cf. Axelrod and Bailey 1969).

Mean annual range of temperature was, during the Oligocene in western Oregon, as great as 21–25°C, but the present value is 12–16°C. At latitude 60° in Alaska, the mean annual range of temperature during the Miocene warm interval was at least as high as 26–27° C, in contrast to the current value of 18°C in the same area (Fig. 4). Similar declines in mean annual range of temperature can be demonstrated in other areas of the Northern Hemisphere, for example, in eastern Asia (Wolfe and Tanai, in press).

Overall trends in mean annual temperature since the terminal Eocene event are dependent on latitude. In southern Alaska (lat 60°), a decline of about 4°C can be documented since the early to middle Miocene. The salient feature of this high-latitude trend is that almost all the change appears to be the result of a decline in summer temperature, which would greatly enhance the "over-summering" of snow fields and, in turn, the initiation of widespread glaciation.

In the Pacific Northwest (lat 42–46° N.), no overall change in mean annual temperature appears to have occurred since the terminal Eocene event. In California and Nevada, climatic inferences from Neogene floras are so greatly complicated by altitudinal and rain shadow factors that extension of these inferences to other areas would at present be unjustified. The few Neogene floras based on leaf remains from eastern North America are too small to be of value in this context.

In Europe, the Neogene floras are found at about the same or higher latitudes as those in the Pacific Northwest; correcting for plate tectonic movements, the Pacific Northwest and European Miocene floras would have been at about equivalent latitudes. The European Neogene sequences typically display—as in the Pacific Northwest—an overall change from broad-leaved deciduous (with a broad-leaved evergreen element, particularly in the Miocene warm interval) to coniferous forest. This implies predominantly a decrease in mean annual range of temperature, possibly along with some decline in warm-month and consequently in mean annual temperatures.

In eastern Asia, the assemblages from Sakhalin (lat 50°) and Kamchatka (lat 55°) show much the same temperature trend as those in Alaska, whereas in Hokkaido (lat 42–45°) only a decrease in mean annual range of temperature occurred (Wolfe and Tanai, in press). South of Hokkaido the floras of Oligocene and Neogene age apparently indicate a contradictory trend—at least in part. In the early Miocene, for example, broad-leaved deciduous forest occupied lowland Kyushu (lat 32°) and southern Honshu (lat 35°; Tanai 1961)—areas now occupied by broad-leaved evergreen forest. Although it can be inferred that mean annual range of temperature has decreased by about 2–4°C, the major point is that mean annual temperature has increased by 3–4°C (Fig. 7). In the middle Miocene of northern Taiwan (lat 25°)—an area now occupied by Paratropical Rain forest—the lowland vegetation was Notophyllous Broad-leaved Evergreen forest (Chaney and Chuang 1968), indicating an increase of at least 2°C in mean annual temperature.

It is significant in this context that Muller (1966) has recorded pollen of elements such as alder and spruce from Borneo (lat 5°), while Graham and Jarzen (1969) have recorded similar cool-climate indicators from the Oligocene to the Miocene in Puerto Rico (lat 18°). In these instances the authors explained the presence of the cool element by suggesting the existence of mountains even higher than those now in the respective areas—although there are no geologic data to support such inferences. Graham (1976), however, explained the presence of cool-climate indicators in the Miocene of Veracruz (lat 19°) by suggesting that temperatures were cooler than now. I suggest that the presence of cool-climate indicators is consistent with the data from Taiwan and Kyushu and implies that, following the terminal Eocene event, low latitudes were cooler than at present.

It is noteworthy that the amount of change in mean annual range of temperature increases as latitude gets higher. At lower latitudes, the major change was apparently an increase in winter temperature that resulted in an overall increase in mean annual temperature and a slight decrease in mean annual range. At about 45° latitude, winter temperature increased by about the same amount as summer temperature decreased, with the result that mean annual temperature remained constant while mean annual range decreased moderately. At high latitudes, summer temperature decreased significantly, leading to a moderate decrease in mean annual range.

One of the obvious consequences of the above trends is an increase in the latitudinal temperature gradient. Such an increase would necessarily increase the intensity of the subtropical high-pressure cells (Willett and Sanders 1959), which, in turn, would bring increasing drought—particularly in summers—to the west coasts of the continents. Such an increase in

Figure 7. It can be inferred from the leaf assemblages that, in southern Japan, broadleaved deciduous forest in the early Miocene gave way to broad-leaved evergreen forest and that mean annual temperature increased by 3–4°C.

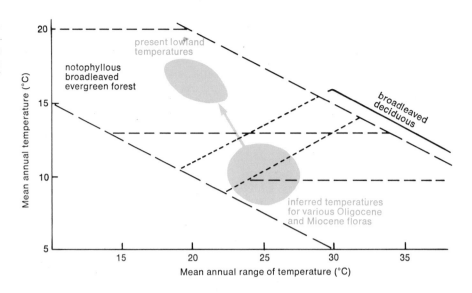

summer drought during the Neogene has been well documented in western North America (Chaney 1944).

Causes of major climatic changes

Milankovitch (1938) proposed that the episodic glaciations of the Quaternary resulted from changes in the inclination of the earth's rotational axis. The changes Milankovitch considered were those due to perturbations, precession, and other known phenomena that result in minor changes in the inclination. In turn, inclinational changes would cause changes in insolation (the amount of radiation received from the sun) that would vary according to latitude. These astronomical explanations of Quaternary climatic change have received considerable support from many researchers, who consider that the paleoclimatic data fit well the timing based on calculations of minor inclinational changes.

Is it entirely coincidental that the divergent latitudinal patterns following the terminal Eocene event fit very well a model resulting from a significant decrease in the inclination of the earth's rotational axis? According to Milankovitch's hypothesis, under conditions of decreasing inclination (and assuming an atmospheric circulation pattern similar to that of the present), the pattern of climatic change would be (1) an increase in winter temperatures at lower latitudes (resulting in an increase in mean annual temperature), (2) an increase in winter temperatures about equal to a decrease in summer temperatures at latitude 43° (resulting in no increase in mean annual temperature), (3) a decrease in summer temperatures at higher latitudes (resulting in a decrease in mean annual temperature), and (4) a decrease in mean annual range of temperature proportional to the latitudes.

These are precisely the changes that are inferred from paleobotanical data

for the Oligocene and Neogene and would indicate that a significant decrease in the earth's inclination has occurred during the last 30 million years.

Conditions during the Paleocene-Eocene were strikingly different from those during the Oligocene and younger epochs. During the thermal and equability maximum of the middle Eocene, the west coast of North America was wet and the southeastern United States was comparatively dry. The presence of humid broad-leaved evergreen forests in Alaska would, concomitant with the other data, argue for a circulation that involved the poleward flow of warm, moist air along the west coast; as this flow returned equatorward over the eastern part of the continent, the heating air would, of course, become drier. That is, the Eocene pattern may have been dominated by north-south (meridional) flow rather than being dominated by regional cells, as at present. What factor(s) could bring about such a pattern?

Readers who are plant physiologists may have been startled to learn that during the Eocene, broad-leaved evergreen forest extended north of latitude 60°. The principles of plant physiology would argue against such vegetation under the prolonged dark winters at such latitudes (Mason 1947; van Steenis 1962). The present distribution of broad-leaved evergreens is consistent with such principles—that is, notophyllous evergreens occur on mountains in California that have the same funda-

mental temperature parameters as lowland areas at more northern latitudes, where notophyllous broad-leaved evergreens are absent (Wolfe, in press). Notophyllous broad-leaved evergreens—except for a very few conspicuous taxa—today do not occur north of 50° latitude, and most are equatorward of 40–45°. Such considerations provide strong evidence that the middle Eocene Alaskan assemblages, with their diverse and dominant notophyllous to mesophyllous broad-leaved evergreen element, could not have existed under the present light conditions at the latitude of the fossil localities. Light was, in fact, considered a negative factor in the possibility of explaining trans-Pacific disjunctions of tropical broad-leaved evergreen groups via the land bridge that is now the Bering Straits (e.g. van Steenis 1962), and yet such groups are now known to have occurred in the Beringian region (van Buesekom 1971; Wolfe 1972).

Under conditions of high temperatures and prolonged winter darkness, the predicted vegetation would be broad-leaved deciduous with a minor broad-leaved evergreen element (van Steenis 1962), composed of those evergreens tolerant of low light levels (as a limited number of broad-leaved evergreen taxa are today). This is apparently the type of vegetation that existed in Alaska north of latitude 60°, during the Paleocene.

I am suggesting that the data thus far indicate that, during the middle Eocene, the light conditions at latitude 60° were more favorable to the growth of broad-leaved evergreens than at present. During the Paleocene, in fact, latitude 57° (but not 60°) supported a diverse broad-leaved evergreen forest. The only factor that could produce more light at these northern latitudes is a significantly smaller (at least 15° less during the middle Eocene than now) inclination of the earth's rotational axis. A smaller inclination would certainly be consistent with the low mean annual range of temperature during the Paleocene and Eocene: today this temperature parameter is primarily (although not entirely) a function of latitudinal position because of the inclination.

A suggestion that the inclination was, in comparison to today, considerably smaller is, if the obvious warmth of the Alaskan middle Eocene is accepted, contradicted by the Milankovitch calculations, i.e. a smaller inclination would yield lower insolation at high latitudes and hence lower temperatures. These calculations were, however, based on the assumption that the insolational values could be directly translated into temperature values—i.e. that the present atmospheric circulation pattern was, in general, constant throughout geologic time. But we have seen that the Paleocene and Eocene circulation pattern could not have been like that of the present. Could an increased insolational gradient under a low inclination have been the major driving force for a dominantly meridional circulation, a circulation that would have more than compensated for decreased annual insolational values at high latitudes? Perhaps at some critical value of inclination, the atmospheric circulation changes from one that is dominantly cellular (as it is today and was during the Oligocene and Neogene) to one that is dominantly meridional.

If the major climatic trends during the Tertiary were largely the result of inclinational changes, then from the Paleocene to the middle Eocene, inclination decreased gradually from a value of perhaps 10° to a value approaching 5°. The inclination then began to increase slightly until the end of the Eocene, when the inclination increased rapidly to 25–30°. Since then, the inclination has gradually decreased to the present average value of 23.5°.

The drastic change in inclination suggested as the cause of the terminal Eocene event would have had a profound effect on the earth's crust. It is significant that a number of researchers have suggested major tectonic changes at the end of the Eocene. For example, Molnar et al. (1975) suggest that the tectonic patterns of the South Pacific were different in the pre-Oligocene than now and that the current patterns were achieved at the end of the Eocene. Menard (1978) similarly indicates major changes in the northeastern Pacific at the end of the Eocene.

Yet, even assuming the validity of this model of inclinational change, the several fluctuations in mean annual temperature are not explained. From available radiometric data, the fluctuations appear to represent a cycle about 9.5 m.y. in duration (Wolfe 1971). Presumably such regular fluctuations would result from fluctuations in the amount of solar radiation reaching the earth; certainly no model of plate tectonic movements could explain such fluctuations.

Much additional information is needed, particularly from the continental interiors, to develop accurate models of temperature and precipitation distribution during the Paleocene and Eocene. Low-latitude leaf floras of Oligocene and Neogene age are needed to determine whether the low-latitude data thus far accumulated are anomalous or typical. More studies of modern depositional environments are needed to understand fully the significance of what is actually found in fossil assemblages. More information from rigidly controlled experiments is needed to determine the physiological response of broad-leaved woody plants to low light levels.

This review has shown the type of data that can be obtained from fossil plants. Brooks (1949) noted that the evidence available to him could not be explained solely by geographic factors, and even attempts to explain Paleocene and Eocene climates by the changing positions of the continental plates are only partially satisfactory (Frakes and Kemp 1973). The evidence now available indicates even more radical differences between present climates and those of the past than were recognized by Brooks. Attempting to fit such data into a "steady-state" hypothesis would be doing an injustice to the data similar to that done to geologic data prior to the general acceptance of plate tectonics.

References

Addicott, W. O. 1969. Tertiary climatic change in the marginal northeast Pacific Ocean. *Science* 165:583–86.

Arnold, C. A. 1955. Tertiary conifers from the Princeton coal field of British Columbia. *Michigan Univ. Mus. Paleontol. Contr.* 12:245–58.

Axelrod, D. I. 1966. The Eocene Copper Basin flora of northeastern Nevada. *Calif. Univ. Pubs. Geol. Sci.* 59:1–125.

Axelrod, D. I., and H. P. Bailey. 1969. Paleotemperature analysis of Tertiary floras. *Palaeogeography, Palaeoclimatology; Palaeoecology* 6:163–95.

Bailey, I. W., and E. W. Sinnott. 1915. A botanical index of Cretaceous and Tertiary climates. *Science* 41:831–34.

Barron, J. A. 1973. Late Miocene-early Pliocene paleotemperatures for California from marine diatom evidence. *Palaeogeography, Palaeoclimatology, Palaeoecology* 14: 277–91.

Berry, E. W. 1914. The Upper Cretaceous and Eocene floras of South Carolina and Georgia. U.S.G.S. Prof. Paper 84.

———. 1916. The Mississippi River bluffs at Columbus and Hickman, Kentucky, and their fossil flora. *U.S. Natl. Mus. Proc.* 48: 293–303.

———. 1929. A revision of the flora of the Latah formation. U.S.G.S. Prof. Paper 154-H, pp. 225–65.

Brooks, C. E. P. 1949. *Climate through the Ages,* 2nd ed. London: Benn.

Chaney, R. W. 1944. Summary and conclusions. In *Pliocene Floras of California and Oregon*, ed. R. W. Chaney, pp. 353–83. Carnegie Inst. Washington publ. 553.

Chaney, R. W., and G. C. Chuang. 1968. An oak-laurel forest in the Miocene of Taiwan (Part 1). *Geol. Soc. China* 11:3–18.

Denton, G., and R. L. Armstrong. 1969. Miocene-Pliocene glaciations in southern Alaska. *Am. J. Sci.* 267:1121–42.

Devereux, I. 1967. Oxygen isotope paleotemperature measurements on New Zealand Tertiary fossils. *New Zealand J. Sci.* 10: 988–1011.

Dilcher, D. L. 1973a. Revision of the Eocene flora of southeastern North America. *Palaeobotanist* 20:7–18.

———. 1973b. A paleoclimatic interpretation of the Eocene floras of southeastern North America. In *Vegetation and Vegetational History of Northern Latin America*, ed. A. Graham, pp. 39–59. Amsterdam: Elsevier.

Frakes, L. A., and E. M. Kemp. 1973. Paleogene continental positions and evolution of climate. In *Implications of Continental Drift to the Earth Sciences*, ed. D. H. Tarling and S. K. Runcorn, pp. 535–58. Academic Press.

Graham, A. 1976. Studies in neotropical paleobotany. II: The Miocene communities of Veracruz, Mexico. *Missouri Bot. Garden Annals* 63:787–842.

Graham, A., and D. M. Jarzen. 1969. Studies in neotropical paleobotany. I: The Oligocene communities of Puerto Rico. *Missouri Bot. Garden Annals* 56:308–57.

Hickey, L. J. 1977. Stratigraphy and paleobotany of the Golden Valley Formation (early Tertiary) of western North Dakota. *Geol. Soc. Am. Mem.* 150.

Jones, D. L., N. J. Silberling, and J. Hillhouse. 1977. Wrangellia—a displaced terrane in northwestern North America. *Can. J. Earth Sci.* 14:2565–77.

Kennett, J. P. 1977. Cenozoic evolution of Antarctic glaciation, the circum-Antarctic ocean, and their impact on global paleoceanography. *J. Geophys. Res.* 82:3843–60.

MacGinitie, H. D. 1953. *Fossil Plants of the Florissant Beds, Colorado.* Carnegie Inst. Washington publ. 599.

———. 1969. The Eocene Green River flora of northwestern Colorado and northeastern Utah. *Calif. Univ. Pubs. Geol. Sci.* 83:1–140.

———. 1974. An early middle Eocene flora from the Yellowstone–Absaroka volcanic province, northwestern Wind River basin, Wyoming. *Calif. Univ. Pubs. Geol. Sci.* 108:1–103.

Mai, D. H. 1964. Die Maxtixioideen-Floren im Tertiär der Oberlausitz. *Palaeontolog. Abh.,* Abt. B, 2:1–92.

Mason, H. L. 1947. Evolution of certain floristic associations in western North America. *Ecol. Monographs* 17:201–10.

Menard, H. W. 1978. Fragmentation of the Farallon plate by pivoting subduction. *J. Geol.* 86:99–110.

Milankovitch, M. 1938. Astronomische Mittel zur Erforschung der erdgeschichtlichen Klimate. *Handbuch der Geophysik* 9: 593–698.

Molnar, P., T. Atwater, J. Mammerick, and S. M. Smith. 1975. Magnetic anomalies, bathymetry, and the tectonic evolution of the South Pacific since the late Cretaceous. *Royal Astron. Soc. Geophys. J.* 40:383–420.

Muller, J. 1966. Montane pollen from the Tertiary of northwestern Borneo. *Blumea* 14:231–35.

Nemjč, F. 1964. Biostratigraphic sequence of floras in the Tertiary of Czechoslovakia. *Časopis pro mineralogii a geologii, Rocnik* 9:107–9.

Plafker, G., and W. O. Addicott. 1976. Glaciomarine deposits of Miocene through Holocene age in the Yakataga Formation along the Gulf of Alaska margin, Alaska. In *Symposium on Recent and Ancient Sedimentary Environments in Alaska,* ed. T. P. Miller, pp. Q1–Q23. Alaska Geological Society.

Raunkiaer, C. 1934. *The Life Forms of Plants and Statistical Plant Geography.* Oxford: Clarendon Press.

Rzedowski, J., and R. McVaugh. 1966. La vegetacion de Nueva Galicia. *Michigan Univ. Herbarium Contr.* 9:1–123.

Spicer, R. A. The sorting of plant remains in a Recent depositional environment. 1975 diss., London Univ.

Tanai, T. 1961. Neogene floral change in Japan. *Hokkaido Univ. Fac. Sci. J.,* ser. 4, 11: 119–298.

———. 1967. On the Hamamelidaceae from the Paleogene of Hokkaido, Japan. *Palaeont.*

Soc. Japan Trans. Proc., N.S., 66:56–62.

———. 1970. The Oligocene floras from the Kushiro coal field, Hokkaido, Japan. *Hokkaido Univ. Fac. Sci. J.,* ser. 4, 14:383–514.

Tanai, T., and K. Huzioka. 1967. Climatic implications of Tertiary floras in Japan. In *Tertiary Correlation and Climatic Changes in the Pacific,* ed. K. Hatai, pp. 89–94. Sendai: Sasaki Printing and Publishing Co.

Triplehorn, D. M., D. L. Turner, and C. W. Naeser. 1977. K-Ar and fission-track dating of ash partings in coal beds from the Kenai Peninsula, Alaska: A revised age for the Homerian Stage-Clamgulchian Stage boundary. *Geol. Soc. Am. Bull.* 88:1156–60.

van Buesekom, C. F. 1971. Revision of *Meliosma* (Sabiaceae) section, *Lorenzanea* excepted, living and fossil, geography and phylogeny. *Blumea* 19:355–529.

van Steenis, C. G. G. J. 1962. The land-bridge theory in botany. *Blumea* 11:235–372.

Webb, L. J. 1959. Physiognomic classification of Australian rain forests. *J. Ecol.* 47:551–70.

Willett, H. C., and F. Sanders. 1959. *Descriptive Meteorology.* Academic Press.

Wolfe, J. A. 1968. Paleogene biostratigraphy of nonmarine rocks in King County, Washington. U.S.G.S. Prof. Paper 571.

———. 1969. Neogene floristic and vegetational history of the Pacific Northwest. *Madroño* 20:83–110.

———. 1971. Tertiary climatic fluctuations and methods of analysis of Tertiary floras. *Palaeogeography, Palaeoclimatology, Palaeoecology* 9:27–57.

———. 1972. An interpretation of Alaskan Tertiary floras. In *Floristics and Paleofloristics of Asia and Eastern North America,* ed. A. Graham, pp. 201–33. Amsterdam: Elsevier.

———. 1977. Paleogene floras from the Gulf of Alaska region. U.S.G.S. Prof. Paper 997.

———. In press. Temperature parameters of humid to mesic forests of eastern Asia and relation to forests of other regions of the Northern Hemisphere and Australasia. U.S.G.S. Prof. Paper 1106.

Wolfe, J. A., and D. M. Hopkins. 1967. Climatic changes recorded by Tertiary land floras in northwestern North America. In *Tertiary Correlation and Climatic Changes in the Pacific,* ed. K. Hatai, pp. 67–76. Pacific Sci. Cong., 11th, Tokyo, Aug. 1966, Symp. 25.

Wolfe, J. A., and T. Tanai. In press. The Miocene Seldovia Point flora from the Kenai Group, Alaska. U.S.G.S. Prof. Paper 1105.

Zhilin, S. G. 1966. A new species of *Carya* from the late Oligocene. *Paleont. Zhur.* 4:104–8 (in Russian). English translation available from Telberg Book Co., New York.

"... and the record low for this date is 147° below zero, which occurred 28,000 years ago during the Great Ice Age."

Margaret B. Davis

Palynology and Environmental History During the Quaternary Period

Dramatic changes in climate have occurred during the last two million years, causing glaciers to form at high latitudes. The glaciers grew in size, spreading out as sheets of ice over the continents, covering at their maximum one-third the land area of the earth. The advance of ice was followed by retreat and readvance, as the climate oscillated between glacial conditions and intervals of warmer climate similar to today. These changes had profound effects upon plants and animals. They caused extinctions, altered distributions, and stimulated rapid evolutionary change.

Even the last few thousand years have witnessed profound change. Only 14,000 years ago tundra and spruce-pine woodland covered much of what is now eastern United States, and stands of spruce forest grew on the high plains. As the climate warmed, and the ice sheet shrank in area, spruce spread into the tundra, growing close to the ice margin in some regions, separated from it by a broad belt of tundra in others. Mammoths and mastodons and caribou grazed on tundra and parkland, and Paleolithic man, newly arrived on the continent, moved across the landscape in their pursuit. But by 9000 years ago there had been a profound change. The tundra disappeared, and spruce was replaced by thick forests of deciduous and coniferous trees. The large herbivores became rare or extinct, and the hunters found new means of livelihood.

These changes in vegetation and climate were discovered by a technique known as palynology. Recognized as one of the most powerful methods in modern paleoecology,

This article emphasizes the vast, dramatic changes in climate during the last two million years and those that have occurred during the last 14,000 years. Palynology is an expanding field in the United States, providing a detailed record of the environment of the past. Margaret B. Davis is Professor and Head of the Department of Ecology and Behavioral Biology at the University of Minnesota. She was educated at Radcliffe College and then studied palynology for a year in Copenhagen as a Fulbright Fellow. Returning to Harvard University, she continued her studies under Hugh M. Raup, obtaining a Ph.D. in biology in 1957. Dr. Davis held successively postdoctoral fellowships at Harvard, California Institute of Technology (in Geology), and at Yale in Zoology. Dr. Davis was formerly on the faculty of the Department of Zoology and the Great Lakes Research Division of The University of Michigan. Address: Dept. of Ecology and Behavioral Biology, University of Minnesota, Minneapolis, MN 55455.

palynology is an expanding field in the United States. A new objectivity characterizes recent work, replacing the intuitive style that long made the method difficult to understand and evaluate. With increasing sophistication, palynology increases its potential for providing a detailed record of the environment of the past. The nature of this record will be described, following an explanation of the way ancient vegetation can be deduced from pollen in sediment.

Pollen in sediment

Palynology is a unique kind of paleontology. Most fossils are the bodies of organisms, each one being evidence of the former existence of an individual plant or animal. Fossil pollen grains, in contrast, are deciduous parts of organisms, produced in large numbers by living individuals, and carried great distances from their source. Their value as fossils stems from their great numbers in sediment. These are quantitatively related, in a complex way, to the vegetation of the surrounding region. Once this complex relationship is understood, the abundances of different plants through time can be deduced from the pollen grains in sediments.

Organic sediments deposited in lakes are often used for pollen analysis. Lake sediment is rich in pollen—a half-million grains per milliliter is not unusual. The grains are, of course, very small, about twenty-five microns in diameter, but their outer waxy coating, resistant to decay, has elaborate morphology, permitting identification of the plant genus, or sometimes even the species, from which the pollen has come. Lakes are found almost everywhere in glaciated regions. Consequently, the paleoecologist interested in the environment at a particular time has little difficulty in finding lakes in the area of interest and recovering from them pollen-bearing sediment of the appropriate age.

Pollen and vegetation

Interpretations of fossil pollen are based on empirically determined relationships between modern pollen and modern vegetation. We can consider an example from Canada. The major vegetation regions there (Rowe, 1959) are shown on Figure 1. In each of these regions samples have been collected from sediments that are accumulating today. A series of collecting sites, along a transect from north to south, are shown by black dots in Figure 1. Each sample was analyzed for its pollen

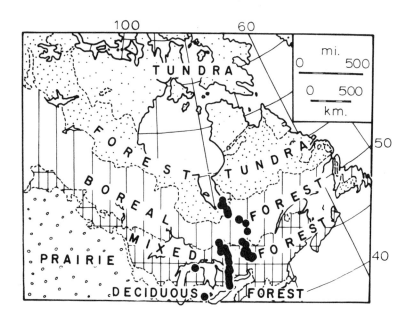

Figure 1. Major vegetation regions in eastern and central Canada modified from Rowe, 1959. The black dots represent sites where surface samples have been collected along a north-south transect.

assemblage, that is, for the types of pollen present, and their relative abundances. These are shown in Figure 2, arranged according to the latitude at which the samples were collected. Clearly each major vegetation region is characterized by a different and distinctive assemblage. Samples from the boreal (coniferous) forest in the north contain high percentages of spruce and pine pollen. Samples from the mixed deciduous-coniferous forest contain high percentages of pine and birch pollen, and those from the deciduous forest at the southern end of the transect are dominated by deciduous tree pollen. Correspondence between pollen assemblage types and vegetation has been found in all parts of the world.

It can be shown at any one site that the percentages of pollen are unequal to the percentages of plants in the vegetation. This is largely because there are differences in the amounts of pollen produced by different species. There are also differences in the ease with which pollen grains of different types are dispersed by the wind, differences in resistance to decay, and so forth. Consequently, the nature of the vegetation cannot be deduced intuitively from the frequencies of pollen in the assemblage. This is so even though geographic trends in modern pollen percentages are often similar (although not precisely parallel) to trends in the abundances of species in the vegetation. But despite the lack of a simple one-to-one correlation between pollen and plant frequencies, the pollen in the sediment is uniquely related to the frequencies of plants. It serves as a signature by which a regional vegetation type can be recognized. In Canada all

the major types of forest that grow today show characteristic assemblages of pollen in surface sediment.

Comparing ancient and modern pollen assemblages

The diagnostic nature of pollen assemblages can be used in the fossil record. If no vegetation change has occurred, then as we sample back in time, the same pollen assemblage found at the surface should be found in deeper, older sediments. But where there have been changes in the vegetation, we will encounter different assemblages in the fossil deposits.

In most regions the fossil pollen shows large changes, indicating that major changes in vegetation did occur in the past. Southern New England is an example. Assemblages dominated by deciduous tree pollen, such as those near the surface, extend back 8000 years (Fig. 3), implying that deciduous forest has grown there throughout this period of time. But in older sediment, deposited 8000-9500 years ago, the assemblage is dominated by pine and birch pollen. It resembles modern pollen assemblages from the mixed coniferous-deciduous forest along the southern margin of the boreal forest in Ontario (Fig. 2). Figure 4 shows the sites where the modern analogs were collected, 800 km north and west from southern New England.

In still older sediment, deposited 10,000 years ago, the assemblages are dominated by spruce and pine pollen

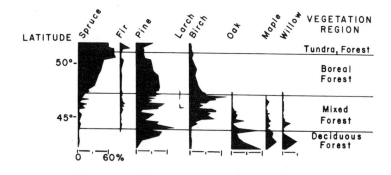

Figure 2. Pollen assemblages in surface samples collected along the north-south transect through Canada shown in Figure 1. The percentage value for each type of pollen is shown on the abscissa. The latitude at which the sample was collected and the forest region are indicated on the ordinate. (After Davis, 1967b.)

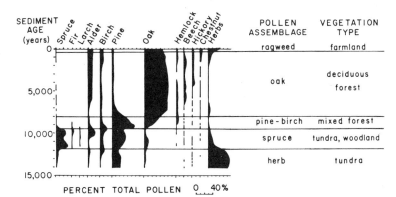

Figure 3. Fossil pollen assemblages in a sediment core from southern New England. The age of the sediment is shown on the ordinate, and the percentages of pollen (somewhat generalized) are plotted on the abscissa. The names of the pollen assemblages and the kinds of vegetation for which they are characteristic are shown to the right of the pollen diagram.

(Fig. 3). They are similar to modern samples collected northeast of the transect shown in Figure 1, at latitude 54° in Quebec north of the boreal forest (Fig. 4). [The detailed comparison is shown in a previous paper (Davis, 1967b)]. The vegetation of this region is "characterized by open, park-like woodlands of black spruce, with an associated ground-cover of lichens" (Terasmae and Mott, 1965, p. 396). Closed forests of black spruce grow there only on wet lowland sites. White spruce, fir, and larch are less common than black spruce, and the nearest jack pine grows about 40 miles away. A similar kind of vegetation, predominantly open spruce-lichen woodland, grew in southern New England ten thousand years ago.

In this way the sequence of fossil pollen assemblages provides a record of vegetation change in southern New England since the end of the Ice Age there. Comparison with modern assemblages shows that within the last 14,000 years the region has changed from (1) tundra to (2) spruce woodland to (3) mixed deciduous-coniferous forest to (4) temperate oak forest. The most dramatic change occurred about 9500 years ago, when the mixed deciduous-coniferous forest replaced subarctic woodland. This occurred without an intervening stage of boreal forest—the broad belt of coniferous forest that to-day separates woodland and mixed deciduous-coniferous forest. With this exception, the succession of forest types through time resembled closely the arrangement of vegetation regions from north to south on the modern landscape of Canada (Fig. 4; see also Davis, 1967b).

Limitations to comparison of pollen assemblages

Comparison of fossil assemblages of pollen grains to modern analogs from suface samples has been remarkably successful, especially in North America. In recent years it has become the most widely used method for interpreting pollen data. But it cannot be used everywhere, because fossil assemblages sometimes occur that have no modern counterpart. The oldest late-glacial sediment in New England, for instance, contains a higher proportion of pollen from temperate trees, such as oak, than modern sediment from tundra regions (4% versus <1%). The oak pollen in modern tundra is transported by wind from distant forests. The higher percentage in the fossil material can be explained in the same way, taking into consideration the geography of the areas occupied by tundra. At present there are vast areas of boreal forest and parkland separating the tundra from the source region for oak pollen. But in the past, without an intervening zone of purely coniferous forest, a mixed forest might have grown closer to the sites of pollen deposition. This source for oak pollen could have provided pollen grains in greater numbers than are found now in tundra assemblages in the high arctic.

Tundra assemblages from the high arctic are easily distinguished from assemblages from forest regions. Near the forest border, however, difficulties arise. Because tundra plants themselves produce so little pollen

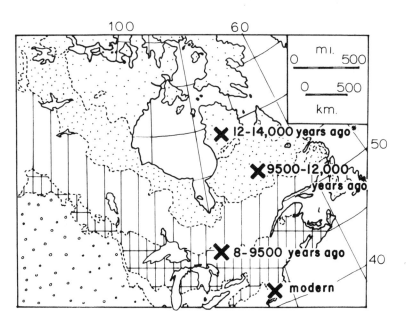

Figure 4. Localities in Canada where surface pollen assemblages resemble fossil assemblages in southern New England. The sites are indicated by x's. The age of the analogous fossil material is indicated for each locality. [*This location in the tundra is extrapolated from the resemblance of fossil material to assemblages from northern-most Quebec (Bartley, 1967). Surface samples are not yet available from the precise locality shown.]

relative to trees, 80 to 90% of the grains in sediment can be tree pollen blown onto the tundra from the nearby forest. Within the forest itself, greater numbers of tree pollen grains are produced. But the herbs within the forest are more productive too, since they are growing in protected sites. For this reason the ratio of tree pollen to herb pollen can be almost identical in tundra near the forest border, and within the forest itself. This makes it difficult to distinguish these vegetation types on the basis of the pollen percentage assemblages they produce (Fig. 5). Fossil pollen deposited 11,000 to 12,000 years ago in New England is subject to this problem. At that time tundra vegetation changed to spruce woodland. Comparison with surface samples is little help in indicating whether at any one point in time the vegetation was predominantly tundra or predominantly woodland. The distinction is important to paleoecologists, since climatic factors have been described in many areas that are correlated with the limit of trees. And the presence or absence of trees is a critical environmental factor affecting the distribution of many animals and plants. But pollen percentages are nearly useless in making the distinction between tundra, parkland, and woodland. To solve the problem an entirely different method of investigation had to be devised.

Pollen accumulation rates

A distinction between tundra and forest can be made by measuring the numbers of pollen grains produced by the vegetation. The numbers of tree pollen grains are different in the two kinds of vegetation, even though the ratio of tree to herb pollen is similar. Recent measurement of the pollen content of the air in the Canadian arctic establishes this point, showing an increase in the numbers of tree pollen grains correlated with an increase in the numbers of trees. The numbers are low in the tundra, increasing in the tundra-forest transition, and higher still inside the forest boundary (Ritchie and Lichti-Federovich, 1967). Therefore, in the fossil record of New England we would expect that the transition from tundra to forest would be marked by a strong increase in the amount of the pollen accumulating in the sediment.

The idea of using the amount of pollen found in sediment to measure the amount of pollen produced by the vegetation was proposed many years ago. But investigators have been frustrated by a lack of time base for the measurement. The concentration of pollen in sediment depends on the number of years represented by each sample, as well as the yearly influx of pollen. Consequently the concentration, without a measurement of time, has no real meaning. But radiocarbon dating has now made it possible to measure the time represented by a unit of sediment thickness at all levels in a sediment profile. With this information, we can use the pollen numbers in a unit volume of sediment to estimate the yearly pollen influx onto the sediment surface. Using this method we have followed quantitatively the changes in yearly tree pollen influx that record the change from tundra to woodland to forest in southern New England (Davis, 1967a, 1969).

The time represented by a unit of sediment thickness was estimated first from a number of radiocarbon dates from various levels within a core of lake sediment. The deposition time for a unit thickness of sediment varied, of course, in different parts of the profile. The amount of

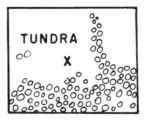
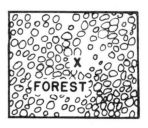

locality A   **locality B**

Figure 5. Schematic distribution of vegetation near the forest-tundra boundary. At locality A, the sampling site (x) is in the tundra, but the majority of the pollen collected is transported from the nearby trees. Similar pollen percentages are found at the sampling site (x) within the forest at locality B.

pollen at each level thus had to be divided by the particular number of years represented by the sample to estimate the influx of pollen per unit surface area each year. The method of calculation is illustrated in Figure 6.

Comparing the yearly influx of pollen to the sediment at various times in the past (Fig. 7), we find that relatively few (600/cm² per year) pollen grains of any kind were deposited 14,000 years ago when there was tundra in southern New England. About 12,000 years ago there was a sharp rise—with the numbers of some kinds of tree pollen increasing 15 to 30-fold. This change marks the arrival of trees in the vicinity. This is corroborated by wood and cones from trees, which first appear in sediment of this age. Although trees were present, they were rare, however. It was some time before spruce woodland dominated the landscape. This is shown by the gradual increase in the annual influx of spruce pollen over the next 2000 years, culminating in maximum rates of accumulation 10,000 years ago (total pollen 10–20,000/cm² per year). Comparison of the pollen percentage assemblage at this level with surface samples has shown that spruce woodland was then the dominant vegetation. The gradual increase in spruce pollen deposition in the sediment below this level records the gradual increase in the numbers of trees on the landscape, as tundra with a few trees gave way first to parkland, and finally to woodland.

A thousand years later, between nine and ten thousand years ago, there was again a sharp rise in the influx of tree pollen, especially white pine, poplar, hemlock, maple, and oak. This increase (total 50,000/cm² per year) reflects the establishment of dense forest. The resemblance of the percentage assemblage at this level to surface samples has indicated the nature of this forest—a mixed deciduous-coniferous community like the modern forests south of the boreal forest region in central Ontario.

Pollen percentage assemblages from certain key levels in the sequence resemble modern assemblages. This has permitted us to identify the vegetation at several points of time in the past. But the transitions between these familiar vegetation types are best understood by observing changes in the amounts of pollen produced by the vegetation. These are recorded by the changes in the amounts of pollen accumulating on the sediment surface.

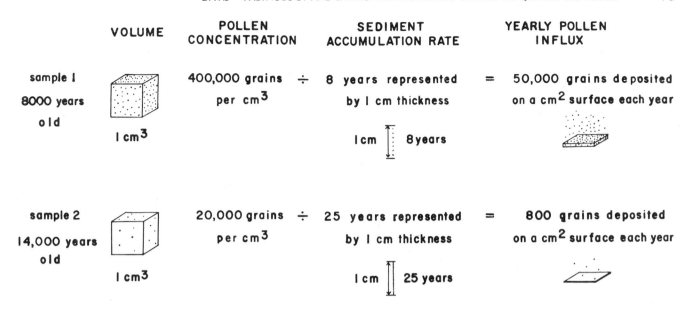

	VOLUME	POLLEN CONCENTRATION	SEDIMENT ACCUMULATION RATE	YEARLY POLLEN INFLUX
sample 1 8000 years old	1 cm^3	400,000 grains per cm^3 ÷	8 years represented by 1 cm thickness 1 cm 8 years	= 50,000 grains deposited on a cm^2 surface each year
sample 2 14,000 years old	1 cm^3	20,000 grains per cm^3 ÷	25 years represented by 1 cm thickness 1 cm 25 years	= 800 grains deposited on a cm^2 surface each year

Figure 6. Comparison of pollen accumulation rates, calculated from the pollen concentration and rate of accumulation of the sediment matrix at two levels in a sediment core.

Direct interpretation of pollen percentages

The nature of ancient vegetation can also be deduced directly from percentages of fossil pollen. This is the most difficult method of interpretation, conceptually and in practice. It was used for many years before the methods described above were developed, and it is still used in those instances were the newer methods cannot be applied. Because the percentages of pollen are quite different from percentages of plants in the vegetation, special methods must be used to arrive at an interpretation. These fall into two major categories.

The first method places emphasis on the changes in percentages from one stratigraphic level to another. The hope has been that an increase or decrease in percentage indicates a similar increase or decrease in the abundance of the parent plant. This works well for most of the last 8000 years in New England, when there was deciduous forest. The total number of pollen grains produced by this forest remained roughly the same for many thousands of years. Consequently changes in the amounts of individual pollen types within the total are reflected accurately by changes in percentage values (compare Figs. 3 and 7); they give a reasonable picture of changes in the composition of the forest. But New England is the only region (as yet) where a uniform annual rate of pollen accumulation has been measured. Until pollen accumulation rates are measured in other parts of North America, and Europe, we cannot be so sure that changes in fossil pollen percentages from one level to another are conveying a true impression of changes in vegetation.

Changes in pollen percentages can be misleading. This is well illustrated by the pine pollen curve in older, late-glacial sediment from New England. The few pine grains blown out onto the tundra from distant forests made up 20% of the pollen total 14,000 years ago. Later a much larger number of pine pollen grains was produced by the woodland and forest vegetation. But since there were also greater numbers of other kinds of pollen grains, the percentage value for pine increases only slightly (Fig. 8). This illustrates how a percentage value is affected by changes in the "background" against which it is projected. Many of the older interpretations of New England vegetation are in error because they failed to take this into account. For example, we supposed at first that the pine pollen maximum 8-9500 years ago represented a forest made up almost exclusively of pine (Deevey, 1939; Dansereau, 1953). Later I suggested the high percentages were an artifact, resulting from low pollen production by the rest of the vegetation (Davis, 1963). The newer data

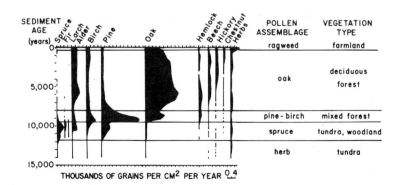

Figure 7. Pollen accumulation rates (somewhat generalized) at a site in southern New England. The age of the sediment is shown on the ordinate. The numbers of grains accumulating yearly on a unit area of sediment surface are plotted on the abscissa. Compare with Figure 3.

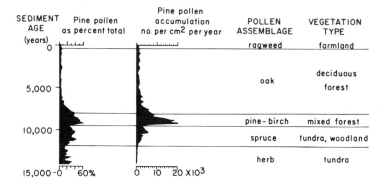

Figure 8. Pine pollen deposition in southern New England sediment. The silhouette on the left indicates the percentage value for pine pollen (as percent total pollen). The number of grains that were accumulating on the sediment each year are shown in the silhouette to the right (after Davis, 1969).

show that both interpretations were incorrect; the number of pine pollen did in fact increase 9500 years ago, but simultaneously with pollen from a number of other tree species, including oak, poplar, maple and hemlock (Fig. 8). These changes record the establishment of a mixed deciduous-coniferous forest, including pine, such as the modern forests of east-central Ontario (Fig. 4).

A second approach to the direct interpretation of pollen percentages corrects for differences between species in pollen production, dispersal, resistance to decay, etc. If compensation is made for these factors, percentages of pollen will equal the percentages of the plants that produced them. The resultant of all the factors affecting representation can be determined empirically in the field. This is done by comparing pollen percentages in surface sediment with abundances of plants in the vegetation. For example, pine and oak might make up 10 and 20% of the pollen, while pine and oak trees make up 5 and 20% of the forest. A point often misunderstood is that studies of this nature measure both plants and pollen as *relative* frequencies. As a consequence, no single species can be considered in isolation. The way each species is represented by pollen must be expressed in comparison to a second species. In this example, representation of pine can be compared to oak: [10% (pollen)] / [5% (plants)] compared to [20% (pollen)] / [20% (plants)]. This reduces to 2 compared to 1, or the ratio 2:1, meaning that pine is twice as well-represented by pollen in proportion to its abundance as oak.

The correction factors (here the ratio 2:1) that result from a study of this kind can be applied to counts of fossil pollen. In this example, if pine and oak were the only pollen types found, the pine counts could be divided by 2 and the oak by 1 before percentages were calculated. The new, corrected percentages should correspond exactly to the percentages of pine and oak trees in the forest.

In practice the method works poorly because an important requirement is seldom met. This requirement is that the representation of each species by pollen, relative to the others, must remain constant through time (pine always twice as well-represented as oak, regardless of abundance). There seems little reason to believe that this is true. Almost everyone who has compared the representation of species in different regions, has found differences in representation from one region to the next (pine 3 or 4 times as well-represented as oak at certain sites).

A few localities have been found, however, where the representation of all the species relative to one another seems constant. These are localities where differences in pollen production are more important than dispersal. A recent study of pollen deposition within a Danish forest, for example, shows little variation in the ratio of representation of the tree species from one sampling site to another (S.T. Andersen, 1967). Accordingly, one can believe that each species was similarly represented by pollen in proportion to its abundance in the past, at least as long as the forest remained of the same general type as today. Here fossil pollen percentages can be corrected to arrive at abundances of parent trees in the ancient forest. But where the vegetation has changed in form, as from tundra to forest, there seems little use in attempting to apply correction factors. Measurement of pollen accumulation rates, or comparison with modern pollen assemblages, are more reliable and useful methods of interpretation.

Potentialities

Influence of man: Palynology was first developed in northern Europe, and it is here that fossil pollen has been studied most intensively. European palynologists have emphasized species identification of fossil pollen, and they have used the ecology of individual species as a basis for interpretation. A lengthy species list is available for the ancient flora. The climate, soils, and even the degree of openness of ancient forests are reconstructed by analogy with the environments of modern populations of these same species of plants.

An intriguing aspect of the European record concerns prehistoric man. Slash-and-burn agriculture by Neolithic people, beginning 5000 years ago, caused characteristic changes in pollen assemblages. These can be traced throughout northern Europe, and even minor differences in agricultural practices can be detected from one region to another. Changes in land use are also recorded in younger sediment, extending into historic time, as the amount of cultivated land increased and new kinds of plants were cultivated. The detail of the record is remarkable. For example, in Denmark a pollen diagram shows a short-lived drop in the amount of agricultural pollen in sediment from the Middle Ages. This was caused by a temporary decline in agriculture, and decrease in the area of cultivated land—a result of devastation by the Black Death during the 14th century (A. Andersen, 1954).

Fossil pollen records the disturbance of European vegetation by man over many thousands of years. The disturbance has been so extensive that its effects cannot always be distinguished from climatic change. There are difficulties in the comparison method of interpretation, too, because pollen assemblages in European surface samples are different from what would be found in regions of undisturbed vegetation. This is why the use of surface samples has developed primarily in North America. On this continent the influence of prehistoric man on the vegetation has been minimal, so far as we know. Even where there has been recent forest clearance for farming, the sediment deposited since then can be detected by its distinctive pollen assemblage (Fig. 3). Samples of the primeval pollen assemblage, deposited under natural conditions, are easily obtained by penetrating beneath this level. Its relative freedom from man's influence makes North America an ideal region for studying biological problems, such as rates of migration, community evolution, and response of vegetation to changes in climate.

Migration of species: The ancient vegetation at any one locality can be described in detail from pollen evidence, as we have shown above for southern New England. When many sites like this have been investigated a number of general statements can be made. The geographical distribution of ancient vegetation regions, for example, can be mapped for a sequence of time intervals. From these maps the migration of plants can be followed in detail: the route followed by each species onto deglaciated territory can be determined. The rate of migration can also be measured using the age differences of successively younger first occurrences in pollen deposits along the migration path. The migration rates can be compared for different species, for different time intervals, for different topography, and against different background vegetation. H. E. Wright, Jr., who has been a leader in this kind of work in North America, has shown that pine was slow to move into areas where it now appears well established (Wright, 1964, 1968). Several deciduous trees, especially beech, hickory, and chestnut, appear to have had a similar history; their late appearance in the pollen record can be seen in Figure 7.

The reasons for delayed migrations and differential migration rates among species are intriguing because they represent a sensitive reaction of a plant population to its environment. Dispersal mechanisms and the availability of migration corridors are only part of the story. The climate and the biotic environment, especially competition from other species, are important also.

Continuity of plant associations: We know already from pollen studies that each plant species has had a different history since the Ice Age. Some boreal trees maintained large populations south of the ice sheet during the last glaciation, while others, presently abundant in Canada, were quite rare. Spruce was common in the Great Plains area, while pine was dominant east of the Appalachians (Wright, 1968; Watts, 1969). Precise reconstructions are difficult because the ancient pollen assemblages are different from any produced by modern vegetation, and cannot be interpreted through comparison. In fact, the comparison method provides a useful means for testing for differences in species composition and frequencies in modern and ancient vegetation. By this criterion some modern plant associations have ancient counterparts, as we observed in tracing the vegetation history of southern New England. Others do not. The modern Boreal Forest formation in Canada, for example, has no clear antece-

dent among the communities south of the ice sheet, nor in those moving northward in the wake of the retreating ice. Despite its great areal extent at present, it appears to be a plant association that was established quite recently. Biogeographic evidence supports this view (Sjörs, 1962). It has no continuity in time and cannot be thought of as the result of strongly-bound adaptations and adjustments of particular species to one another. It must be explained as a recent adjustment in a particular environment rather than the product of community evolution during a long and continuous history.

The differences between ancient and modern plant associations have a further implication. The vegetational environment faced by animals that had moved southward ahead of the advancing ice sheet was different from any existing today. Many reconstructions of environments during the Ice Age have assumed an overly simple southward displacement of all existing biotic zones. These must be reexamined, in the light of growing evidence, for a different distribution and abundance of many plant species south of the ice sheet than occur in arctic and boreal regions today.

Paleoenvironmental reconstructions: The power of palynology as a method for investigating past environments stems from its ability to identify the dominant vegetation on the landscape. This implies something about the regional climate, even without knowledge of the critical climatic factors affecting vegetation. The geographical location of modern pollen assemblages analogous to ancient ones expresses this implication, giving by analogy a general idea of ancient climate. But to carry the paleoclimatic inferences to a more precise level, critical factors controlling plant abundances will have to be identified and understood. This is the area of research that must be emphasized for further progress. We need to know the effect of climate, and biotic factors, on the survivorship of marginal populations of migrating plant species. We need to know more about the dynamics of plant community structure to understand the effect of the removal or addition of a dominant species. Fossil pollen provides a record of changing population size; if this is to be interpreted in a sophisticated way, we need to know more about the population biology of the species concerned.

Summary

Pollen grains preserved in sediment provide a record of the terrestrial vegetation surrounding a site of deposition. In southern New England after the retreat of the ice sheet, for example, tundra changed first to park-tundra and then to spruce woodland. Before a closed boreal forest could develop on the landscape, conditions changed to favor a more temperate kind of vegetation. This was a mixed coniferous-deciduous forest like the modern forests of east-central Ontario. Later a temperate deciduous forest became established. With similar information from widely-distributed sites, vegetation changes can be compared over large regions, and the migrations of species can be followed as they expanded northward onto deglaciated landscape. Modern communities can be compared with ancient ones; some plant associations appear to have had a long history, while others are the product of recent adjustments to the postglacial environment. With sufficient geographical coverage we shall be able to evaluate the effect regional vegetation change has had on animal populations, to assess its influence on high extinction rates of mammals, on rapid rates of morphological evolution, and on rapid development of human culture that characterize the end of the Ice Age.

Acknowledgments

This paper was written with the support of the National Science Foundation, Project grant GB 7727. I am grateful to the following for critical reading of various drafts of the manuscript: W. Charles Kerfoot, Linda B. Brubaker, Jane M. Beiswenger, W. T. Edmondson, and H. E. Wright, Jr. The paper is Contribution No. 107 from the Great Lakes Research Division, University of Michigan.

References

Andersen, A. 1954. Two standard pollen diagrams from south Jutland. *Danmarks geol. Unders,* II Series, No. 80, pp. 188–209.

Andersen, S. T. 1967. Tree-pollen rain in a mixed deciduous forest in south Jutland (Denmark); *Rev. Palaeobotan. Palynol., 3,* p. 267–275.

Bartley, D. D. 1967. Pollen analysis of surface samples of vegetation from arctic Quebec. *Pollen et Spores, 9,* p. 101–105.

Dansereau, P. 1953. The postglacial pine period. *Roy. Soc. Can. Trans.,* III, Sec. 5, *47,* p. 23–38.

Davis, M. B. 1963. On the theory of pollen analysis. *Amer. Jour. Science, 261,* p. 897–912.

——. 1967a. Pollen accumulation rates at Rogers Lake, Connecticut, during late- and postglacial time. *Rev. Palaeobotan. Palynol., 2,* p. 219–230.

——. 1967b. Late-glacial climate in northern United States: a comparison of New England and the Great Lakes region. p. 11–43, *in* E. J. Cushing and H. E. Wright, Jr., *Quaternary Paleoecology*, New Haven, Yale Univ. Press, 433 p.

——. 1969. Climatic change in southern Connecticut recorded by pollen deposition at Rogers Lake. *Ecology (in press).*

Deevey, E. A., Jr. 1939. Studies on Connecticut lake sediments. I. A postglacial climatic chronology for southern New England. *Amer. J. Sci., 237,* p. 691–724.

Ritchie, J. C., and Lichti-Federovich, S. 1967. Pollen dispersal phenomena in arctic-subarctic Canada. *Rev. Palaeobot. Palynol., 3,* p. 255–266.

Rowe, J. S. 1959. Forest regions of Canada. Canadian Dept. Northern Affairs and Natural Resources, Forestry Branch, Bull. 123, 71 p.

Sjörs, H. 1963. Amphi-Atlantic zonation, Nemoral to Arctic. p. 109–126 *in* A. Love and Love, *ed., North Atlantic biota and their history,* Oxford, England, Pergamon Press, 430 p.

Terasmae, J., and Mott, R. J. 1965. Modem pollen deposition in the Nichicun lake area, Quebec. *Canad. J. Bot., 43,* p. 393–404.

Watts, W. A. 1969. The full-glacial vegetation of northwestern Georgia. *Ecology (in press).*

Wright, H. E., Jr. 1964. Aspects of the early postglacial forest succession in the Great Lakes region. *Ecology, 45,* p. 429–448.

——. 1968. The role of pine and spruce in the forest history of Minnesota and adjacent areas. *Ecology, 49,* p. 937–955.

PART 3 *Influence of Climate on People and People on Climate*

Daniel A. Livingstone

Speculations on the Climatic History of Mankind

Inference from the fossil record of lake deposits suggests how climate may have controlled the distribution and abundance of man over the ages

A few mammals, such as the cougar, the leopard, and the Norway rat, are remarkable for their wide geographic range. They exploit steppe and rain forest, they tolerate the full range from tropic heat to arctic cold. Their ubiquity does not invite climatic speculation. An ecologist will explain their wide ranges in terms of adaptability, heterozygosity, or generalized food habits rather than isotherms of maximum summer temperature or isopleths of winter snowfall. In this small group of cosmopolites no species is so wide-ranging as our own, and we might seem the world's poorest mammalian material for biometeorological investigation. Add to this the poverty of the human fossil record, and the climatic history of mankind begins to seem speculative indeed.

We persist in the pursuit of so unpromising a subject only partly because we are men ourselves. Two things make climate and man a fruitful symposium subject. First, there is our common experience that the abundance and activity of man, if not his geographical range, are under continual climatic restraint. Even the vagaries of weather chafe and bind us. Second, even such climatic independence as we now enjoy is a very recent benefit of our cultural evolution. Through most of human history our attainments were more modest and our geographic distribution under much stricter climatic control.

Daniel A. Livingstone, Professor of Zoology at Duke University, received his training in biology at McGill, Dalhousie, Yale, and Cambridge. He has worked in chemical embryology, fisheries biology, geochemistry, paleolimnology and Pleistocene geology, but is known chiefly for having been tasted and rejected by a large Nile crocodile. The general line of investigation reported here has been supported by the Guggenheim Foundation and several grants from the National Science Foundation to Duke University. The paper is modified from one presented at an American Meteorological Society symposium on Climate and Man that was held at Boston in 1969. The author wishes to acknowledge the aid of his colleagues, teachers, and friends M.B. Davis, E. S. Deevey, Jr., J. B. Gillett, F. R. Hayes, M. J. Heck, G. E. Hutchinson, R. L. Kendall, J. L. Richardson, and the late H. P. Bell at various stages from initial inspiration to preparation of the final typescript. Address: Zoology Department, Duke University, Durham, NC 27706

Without fire, lamps, warm clothing, houses, or snowshoes men could occupy but a small part of the enormous area in which they are found today.

Climates of the past

Probably the richest store of pertinent information about past climate lies in lake mud. In a desert the very existence of lake mud is a valuable climatic indicator. Gypsum or halite beds in the mud indicate prolonged evaporative concentration of the lake water (1). Raised shorelines high above the water level of existing lakes, or depositional hiatuses far beneath the present water level, where the earth's crust has been stable, indicate moister and drier episodes, respectively (2).

Human relics of the Lower and Middle Paleolithic period are so commonly associated with rivers and with lakes as to suggest that the people of those times had no effective way of carrying fluids and were obliged to pass their lives within easy reach of potable water. If so, the existence and salinity of lakes assume a special importance for the history of man, quite apart from their general value as indicators of past climate.

Lake sediments contain many indicators of past climate, but I will dwell on only one, the fossil pollen grains. Pollen grains and spores are well preserved in lake mud, especially if it has been acid and moist continuously since its formation. Ordinary organic lake mud commonly contains some hundreds of thousands of these microfossils per milliliter, and it is a simple technical task to separate them from their matrix for microscopic examination.

The ease and accuracy with which pollen grains can be identified varies from family to family, but accurate identification to genus can be made by light microscopy for most of the flowering plants. Electron microscopy is useful in unraveling the structure of the pollen wall, but less so in fossil identification because the grains are too electron-dense for direct observation. Scanning electron microscopy, which provides images of the surface of intact objects, may permit accurate identifications to species in some groups of plants, but is not yet in common use among pollen analysts (3).

Vegetational conclusions from pollen analysis are less certain than the pollen identifications. Plants differ greatly in the amount of pollen they produce and the distance to which they distribute it. Wind-pollinated species broadcast large quantities of pollen and are best represented in the fossil record. Species pollinated by insects, birds, and bats are represented poorly or not at all. In the temperate zone, where most forest trees and many herbs and shrubs are wind-pollinated, this is not a serious limitation, but in some kinds of tropical vegetation dominated by insect-pollinated legumes it may obscure the vegetational history.

The conclusions of pollen analysis are primarily vegetational ones. Vegetation is a function of climate, to be sure, but climatic conclusions depend on additional inference and are correspondingly less secure. This is a serious disadvantage if one is primarily concerned with paleoclimatology, but for our purpose it is less severe. We are interested in the environment of man as affected by changing climate, and an accurate record of past vegetation is probably a more valuable datum for that purpose than a set of thermometer readings.

For example, we may be interested in the conditions early man met while entering America via the Bering Strait. The modern climate around the Strait is subarctic to arctic, and no amount of conditioning and fortitude would permit a man to survive its winter without warm clothing. It is conceivable, though, that there may have been a time when a friendlier climate would permit hunters to filter across from Asia before they had attained any great skill in the dressing of skins and the manufacture of fur clothing. Some local vagary in interglacial of interstadial warming might have moderated the Bering climate while the Bering platform remained above sea level, or tectonic activity around the Bering Strait might have created a land bridge during warm interglacial time.

Colinvaux (4) has used pollen analysis to show that conditions on Seward Peninsula and the islands of the Bering Sea have not been milder than they are now for a very long time. During ice ages the vegetation consisted of tundra comparable to that growing much farther north in modern Alaska.

During interstadial, interglacial, and postglacial times the vegetation was like that of today. People unequipped for a tundra winter cannot have crossed the Bering land bridge during a time that extends back at least to the middle of the last interglacial period.

Man's arrival in the New World, on present evidence, appears to have occurred about 13,000 years ago. This was a time when much of North America, including possibly the route into the continent, was under glacier ice. If ice really did block the way men must have skirted it in boats. Archaeologists are reluctant to concede the existence of seaworthy craft at a date so much earlier than their appearance in the record, but the level of technical proficiency required for the manufacture of tailored skin clothing is not markedly different from that required to build a curragh or umiak.

Whatever the details, the technical development that permitted men to penetrate the Bering climatic barrier was a long step toward climatic independence. It opened up North and South America for settlement, most of which would have been inhabitable at a lower cultural level had primitive men been able to reach it. Mastery of the Bering environment implies habitation of Siberia and northern Europe as well, territories that could not have been occupied by a more climate-bound people. Even without the peopling of Australia, which seems to have occurred at about the same time and to have depended upon the combination of some rudimentary seamanship and a climatically lowered sea level, this late Paleolithic revolution stands as a unique event in the territorial expansion of man.

The coldest environment that has ever been inhabited for an extended period of time is Peary Land, north of the Greenland ice sheet. Apparently the period of prehistoric Eskimo occupation coincided with relatively mild conditions (5). Much has been made of possible climatic controls on the medieval Norse settlement in the much milder climate of southwest Greenland. In the lake sediments of Greenland, however, it is easier to discern the influence of man on the vegetation than the influence of climate and vegetation on man (6).

Neolithic cultural remains are found in parts of the Sahara now too dry to support such a culture, and a growing body of geological evidence (7) shows that the desert has been less dry in the relatively recent past. No detailed study seems to have been made of archaeologic and paleoclimatic evidence in a place where both are abundant. Perhaps the most promising place for such work is Jebel

Marra (8), a massif rising out of the Sudan, which is surrounded by indications of former agriculture, bears remnants of an olive forest that would register well in the pollen record, and possesses lakes to preserve that record.

The climatic environment of people occupying such an isolated habitat is easier to specify than that of people whose habitat is extensive and diverse. In western Iran, for example, van Zeist (9) has shown that *Artemisia* steppe gave way to oak-pistachio savannah some 10,000 years ago. This is about the time cave-dwelling gave way to village life in the region, starting a chain of events that has led to our present civilization. As Wright (10) points out, however, this is an area of considerable relief, with a wide variety of vegetation types persisting at various altitudes into the present day. We can hardly think that cave-dwellers, pinched by a disappearing habitat, would invent a fundamentally new way of living if they had only to move their encampments a few hundred meters down the hillsides to follow the retreating steppe. The coincidence of climatic and cultural change is a striking one, nevertheless, particularly when it is viewed in broader perspective. The climatic change of 10,000–12,000 years ago appears to have been worldwide, and the cultural events associated with it in Kurdistan are among the most important in human prehistory.

Climatic change in the tropics

Our unclothed thermal tolerance and our inability to synthesize vitamin D except in sunlight, or to synthesize ascorbic acid at all, suggest that we are, primevally, creatures of a tropical environment. So does the distribution of our parasitic diseases, the number of which in tropical regions suggests a long period of evolutionary adaptation to a human host. More convincingly, the distribution of old and primitive hominid fossils, of our close cousins the gorilla and chimpanzee, and our more distant cousins the orangutan and gibbon, all suggest a tropical home for the human stock. Let us consider the climatic nature of the tropics, and particularly ways in which the tropical climate has changed to affect man.

In the temperate zone the most dramatic evidence of climatic change comes from continental ice sheets. Most of the tropical zone has been too warm for glacier growth during Cenozoic time, and the lack of extensive till sheets has sometimes misled people into viewing equatorial regions as places of constant climate. Actually the tropical climate has changed greatly, particularly with respect to humidity, during even the past few thousand years. In Africa and Australia raised strandlines indicate at least one former period when lakes were much deeper, or when there were lakes in basins that are now quite dry. In equatorial West Africa extensive areas are underlain by wind-blown sands where precipitation is presently much too high for drifting sand.

Tropical mountains that are high and wet enough to maintain glaciers now, and some from which all vestiges of permanent ice have disappeared, are ringed by terminal moraines at altitudes that indicate a firn line, or limit of net snow accumulation, some thousand meters below the modern one (11). Such an ice advance implies temperature lowering of 4°C with the present precipitation regime, or more if the climate was dry as well.

Former glaciers, landslides, and volcanoes have all produced lakes on tropical mountains, and the sediments of these lakes have been analyzed in Hawaii (12), New Guinea (13), Costa Rica (14), Colombia (15), and the mountains of East Africa (16). In all of these places the pollen spectra show changes during the past ten or twenty thousand years comparable with those of the temperate zone. All investigators are agreed that the broad features of climatic change have been similar at temperate and tropical latitudes: a major ice retreat began some twelve to fifteen thousand years ago. There are differences of opinion about the details of climatic and vegetational changes since then.

Kendall (17) showed that the vegetation around Lake Victoria in Uganda had also gone through profound changes during the past fifteen thousand years. His pollen analyses were combined with chemical studies showing that the lake lacked an outlet and was surrounded by savannah during the closing phase of the last glaciation. The water level rose as the vegetation changed to tropical rain forest about 12,000 years ago.

Carbon dates of raised beaches in East Africa (2, 18) support Kendall's thesis that the last pluvial period in East Africa occurred during postglacial time, and results from the Chad basin (7) indicate the same thing. Longer dated sections are needed to establish the generality of the conclusion, but all of this evidence suggests that tropical lowlands were wet during the postglacial period and dry during the last glaciation. If this correspondence of glacial and interpluvial periods is generally true, the tropics have been subjected to a series of alternating wet and dry periods during Pleistocene time.

In climate and vegetation East Africa is very different from the sultry jungle immortalized by Conrad in *Heart of Darkness*. Lying for the most part between one and two thousand meters, the East African plateau displays many local variations of climate and vegetation where faulting and vulcanism have given grain to its topography. In general, though, the climate is dry and not oppressively hot, and the vegetation is open woodland or savannah. We cannot afford to pass it by for the more typical equatorial conditions of Amazonia or the Congo basin, however, because this anomalously dry and cool region has been the home of man for a long time. It shares with adjacent parts of southern Africa the richest fossil record, covering the longest span of time. It may be the ancestral home of humanity. The long unbroken sequence of cultural changes uncovered by L.S.B. and Mary Leakey at Olduvai Gorge in Tanzania contrasts sharply with the abrupt hiatuses in the Paleolithic record of Europe and gives a more positive endorsement of the East African claim to be the home of human culture.

African lakes are very old and may ultimately present us with a vegetational record covering a large part of the Neocene. Until the technical problems of raising complete sections from deep mud deposits under deep water have been solved, however, we are restricted to a much shorter span of time—in lowland localities, to the 15,000 years studied by Kendall.

His record shows former extensions of grass-dominated vegetation and of rain forest over an area that is now a forest–savannah mosaic. These are major shifts in vegetation, and they imply major shifts in climate. For example, his work and current studies by Thomas Harvey (personal communication) indicate that the entire Upper Nile basin was an area of internal drainage as recently as 13,000 years ago. Such shifts in vegetation and climate, however, need not imply a fundamental change in the nature of the East African environment. I have given reasons elsewhere for my belief in the mutability of plant communities (*16*) and do not wish to imply that existing associations of species merely shifted their borders slightly in response to climatic change. A variety of woodland, grassland, and savannah types occupied most of the landscape through Pleistocene time. Only the proportions of these sorts of vegetation, and the area under rain forest, changed with the climate.

One can be less assured of the continued presence of evergreen forest vegetation. It is of limited extent

today, was much scarcer prior to 12,000 years ago, and may well have vanished completely during earlier and more severe dry periods. A forest tribe of hunter-gatherers like the modern Bambuti would be seriously affected by a decrease in rain forest and might have disappeared from East Africa during dry times. The archaeological record suggests, however, that such forest dependence is a relatively recent phenomenon. It is only with the development of the Sangoan culture within the past 40,000 years that signs of a forest way of life become evident (*19*). In earlier times, and this includes the overwhelming part of our history, we seem to have possessed only a very general material culture, well adapted to a variety of the more open vegetations and dry climates to which our upright gait, our eye-dominated sensory system, and our evaporative thermoregulation appear to suit us. This sort of environment certainly became more and less abundant in response to climatic change, and the location as well as the sizes of the separate patches of which it was composed certainly shifted, but there seems no reason to doubt that East Africa contained, throughout the Pleistocene, large savannah areas suitable for habitation by early man.

Europe and Asia

I have said little about the continents that seemed forty years ago to be the main theater of human evolutionary events, because most of the things that might be said are old and familiar, or because the facts on which new statements would have to rest are not available. This, regrettably, is the case for the details of cultural development that permitted late Paleolithic men to advance through the cold parts of Asia to the Bering bridge. The organic nature of most of the equipment needed to cope with the sub-zero winters offers little hope that we will ever have a satisfactory understanding of that particular revolution, but careful study of living sites and a few happy accidents of preservation might produce a surprising amount of insight that we now lack.

No doubt the northern limit of lower and middle Paleolithic settlement in temperate Eurasia retreated during each ice age, but it does not follow that the world population of men was maximal under interglacial conditions. The interglacial vegetation of Western Europe, for example, was primarily forest, although there were places and times, as West (*20*) has shown, where a more open vegetation prevailed. Forest is not good territory for primitive hunters. The visibility is poor, much of the mineral nutrient store is locked up in wood, and large parts of its primary production are likely to be funneled into animals such as birds and insects that are less easily harvested than the large grazing mammals of open country. In Africa, at least, forest living sites were very rare until late Paleolithic time (21).

Application of pollen analysis to Mediterranean Europe and the Middle East has shown that the glacial vegetation was primarily steppe, with trees being much less important than they were in postglacial time (9, 22). There is some suggestion that ice ages were marked by moister conditions in the northern Sahara (23) and Nubia (24), as they certainly were in the now arid southwestern United States. We have seen that the last ice age seemed to favor savannah in equatorial Uganda. In all of these areas one would expect the greatest human population during ice ages. The southern Sahara, however, would be most suitable for man during the moist, pluvial conditions of postglacial or, presumably, interglacial time. Western Europe would probably maintain its greatest human density during the late glacial phase of each ice age, when the climate had warmed to something approaching interglacial temperatures, but when dense forest had not yet spread over the land.

If man evolved in East or southern Africa and then spread north to temperate Europe, there would be clear advantages to his riding the wave of environmental changes associated with an ice age. He could move into the southern Sahara during interglacial time, move on to the northern Sahara and the Mediterranean during the next glaciation, and advance north of the Alps and Pyrenees in the ensuing late glacial period.

The arid Middle East connecting Asia and Africa is a climatic barrier separating two more favorable areas, very much as the cold Bering climate separates Asia and America. There are enough similarities in the mammalian faunas of India and Africa to show that this gap has been crossed occasionally, but enough differences to show that it has generally been a formidable barrier. Perhaps the initial bridging of the gap by men or their ancestors was, like the peopling of the New World, aided by cultural developments.

Adaptation to climate

Direct relations between climate and the history of early man are difficult to demonstrate. The time of rapid climatic change ten to fifteen thousand years ago was one of rapid cultural development also. Settlement of the New World, development of highly specialized microlithic cultures, domestica- tion of plants and animals, and the adoption of settled village life all seem to have occurred during that span of time. On the other hand, writing, cities, and smelting began under conditions of relative climatic calm. Cultural development is obviously an autocatalytic reaction, still in something close to its exponential phase. Any reasonably long span of recent time, chosen quite at random, is likely to include a certain number of cultural innovations important for their immediate impact or for their ultimate consequences. Whether that particular five-thousand-year span of time saw more innovations than one would expect on a random basis depends partly on one's subjective judgment of the importance of the particular innovations themselves.

The common response of animals to environmental change is genetic adaptation. In a heterozygous stock the selection can occur quite fast, within a few generations, as users of insecticides and antibiotics know to their sorrow. In a homozygous stock it may take much longer, but even so, genetic response would be quite adequate for meeting the slow environmental changes of early Cenozoic time.

During latter Neocene time, and especially the Pleistocene, environmental change appears to have accelerated, placing a premium on adaptability. Our brains permit cultural adaptation that is faster than genetic selection. This gives us a competitive advantage over mammals of comparable generation time that depend on genetic adaptation, and it may have been a factor in our great Pleistocene success.

Not all animals have so long a genetic response time, however. Insects with a generation time of only a week or two can adapt their genome much more rapidly than we can change our culture. In recent years human culture itself seems to change our environment too rapidly for an easy cultural response. By the large-scale use of energy, by pollution, and by the free use of persistent poisons, we are altering the prevailing rate of environmental change from one we handle easily to one insects can handle more easily than ourselves.

References

1. Bradley, W. H., and H. P. Eugster. 1969. Geochemistry and paleolimnology of the trona deposits and associated authigenic minerals of the Green River formation of Wyoming. U.S. Geol. Surv. Prof. Paper 496-B, 1–75.

2. Washbourn, C. K. 1967. Lake levels and Quaternary climates in the Eastern Rift Valley of Kenya. *Nature*, 216:672–73. Washbourn-Kamau, C. K. 1970. Late Quaternary chronology of the Nakuru—Elmenteita basin, Kenya. *Nature* 226:253–54. Richardson, J. L. 1969. Former lake-level

fluctuations—their recognition and interpretation. *Mitt. int. Verein. theor. angew. Limnol.* 17:78–93.

3. Martin, P. S., and C. M. Drew. 1969. Scanning electron micrographs of Southwestern pollen grains. *Jour. Arizona Acad. Sci.* 5:147–76.

4. Colinvaux, P. A. 1967. Quaternary vegetational history of arctic Alaska. p. 207–31 in *The Bering Land Bridge*, D. M. Hopkins, ed. Stanford Univ. Press, Stanford, Cal.

5. Fredskild, Bent. 1969. A Postglacial standard diagram from Peary Land, North Greenland (1). *Pollen et Spores*, Vol. XI, No. 3, 573–83.

6. Iversen, J. 1952–3. Origin of the flora of western Greenland in the light of pollen analysis. *Oikos* 4:85–103.

7. Faure, H. 1969. Lacs Quaternaires du Sahara. *Mitt. int. Verein. theor. angew. Limnol.* 17:131–46.

8. Lebon, J. H. G., and V. C. Robertson. 1961. The Jebel Marra, Darfur, and its region. *Geogr. Jour.* 127:30–49.

9. van Zeist, W. 1967. Late Quaternary vegetation history of western Iran. *Rev. Palaeobotan. Palynol.* 2:301–11.

10. Wright, H. E. 1964. Late Quaternary climates and early man in the mountains of Kurdistan. *Rept. VIth. Int. Congr. Quat. Vol. II: Palaeoclimatol. Sect.:* 341–48.

11. Osmaston, H. A. 1965. The past and present climate and vegetation of Ruwenzori and its neighbourhood. D. Phil. thesis, Worcester Coll., Oxford, n.p.

12. Selling, O. H. 1948. *Studies on Hawaiian pollen statistics, III. On the late Quaternary history of the Hawaiian vegetation.* B. P. Bishop Mus. Spec. Pub. 39, 1–119.

13. Flenley, J. R. 1967. The present and former vegetation of the Wabag region of New Guinea. Ph.D. thesis, Austral. Nat. Univ., n.p.

14. Martin, P. S. 1964. Paleoclimatology and a tropical pollen profile. *Rept. VIth. Congr. Quat. Vol. II: Palaeoclimatol. Sect.:* 319–23.

15. van der Hammen, T., and E. Gonzalez. 1960. Upper Pleistocene and Holocene climate and vegetation of the "Sabana de Bogota" (Colombia, South America). *Leidse Geologische Mededlingen* 25:261–315.

16. Coetzee, J. A. 1967. Pollen analytical studies in East and Southern Africa. *Palaeoecology of Africa*, III. E. M. van Zinderen Bakker, ed.: 1–146. Bakker, E. M. van Zinderen. 1964. A pollen diagram from equatorial Africa, Cherangani, Kenya. *Geol. en Mijnbouw* 43:123–28. Livingstone, D. A. 1967. Postglacial vegetation of the Ruwenzori Mountains in equatorial Africa. *Ecol. Monogr.* 37:25–52.

17. Kendall, R. L. 1969. An ecological history of the Lake Victoria basin. *Ecol. Monogr.* 39:121–76.

18. Livingstone, D. A., and R. L. Kendall. 1969. Stratigraphic studies of East African lakes. *Mitt. Internat. Verein. theor. angew. Limnol.* 17:147–53. Announcement by Glynn Isaac at Fifth African Prehistory Congress, Teneriffe, 1963.

19. Clark, J. D. 1965. Changing trends and developing values in African prehistory. *African Affairs*, special spring number, n.p.

20. West, R. G. 1956. The Quaternary deposits at Hoxne, Suffolk. *Phil. Trans. Roy. Soc. Lond.* Ser. B. 239:265–356.

21. Clark, J. D. 1965. Culture and ecology in prehistoric Africa. p. 13–28, in D. Brokensha, ed., *Ecology and Economic Development in Tropical Africa.* Inst. Int. Stud., Univ. Cal., Berkeley.

22. Bonatti, E. 1961. I sedimenti del lago di Monterosi. *Experientia.* 17:252–53. Bonatti, E. 1966. North Mediterranean climate during the last Würm glaciation. *Nature* 5027:984–85. van der Hammen, T., T. A. Wijmstra, and W. H. van der Molen. 1965. Palynological study of a very thick peat section in Greece, and the Würm-Glacial vegetation in the Mediterranean region. *Geologie en Mijnbouw.* 44:37–39.

23. Alimen, H., F. Beucher, and G. Conrad. 1966. Chronologie du dernier cycle Pluvial-Arid au Sahara nord-occidental. *C.R. Acad. Sci. Paris* 263:5–8. Conrad, G. 1969. L'Evolution continentale Post-Hercynienne du Sahara Algerien. Centre de Recherches sur les Zones Arides. Serie: Geologie 10: 527 p.

24. Butzer, K. W., and C. L. Hansen. 1968. *Desert and River in Nubia.* Univ. Wisconsin Press, 562 p.

C. F. Baes, Jr., H. E.
Goeller, J. S. Olson, and
R. M. Rotty

Carbon Dioxide and Climate: The Uncontrolled Experiment

Possibly severe consequences of growing CO_2 releases from fossil fuels require a much better understanding of the carbon cycle, climate change, and the resulting impacts

According to Revelle and Suess (1957), "Human beings are now carrying out a large-scale geophysical experiment of a kind that could not have happened in the past nor be repeated in the future. Within a few centuries we are returning to the atmosphere and oceans the concentrated organic carbon stored in the sedimentary rocks over hundreds of millions of years. This experiment, if adequately documented, may yield a far-reaching insight into the processes determining weather and climate." Thus well said is the need to *observe* vigilantly the consequences of man's consumption of fossil fuels—coal, oil, and natural gas—and the concomitant return of vast amounts of carbon to the atmosphere in the form of carbon dioxide.

Left unstated, however, is perhaps

C. F. Baes, H. E. Goeller, and J. S. Olson are, respectively, members of the Chemistry Division, the Program Planning and Analysis Office, and the Environmental Sciences Division of Oak Ridge National Laboratory. R. M. Rotty is a member of the Institute for Energy Analysis, Oak Ridge Associated Universities. The paper is an adaption of a report entitled "The Global Carbon Dioxide Problem" (U.S. Energy Research and Development Report ORNL-5194). The research was sponsored by the Energy Research and Development Administration under contract with Union Carbide Corporation. During the course of this review, the authors have benefited greatly from discussions with a number of individuals; they wish especially to thank W. Broecker, of the Lamont-Doherty Geological Observatory; V. Ramanathen, of the National Center for Atmospheric Research; L. Machta, Director of the Air Resources Laboratories, NOAA; J. M. Mitchell, of the Environmental Data Service, NOAA; S. Manabe, of the Geophysical Fluid Dynamics Laboratory, NOAA; and G. Marland, Oak Ridge Associated Universities. Address: Oak Ridge National Laboratory, or, for R. M. Rotty, Oak Ridge Associated Universities, Oak Ridge, TN 37830.

the greater need to *anticipate* the consequences of this process well enough in advance to keep them within acceptable limits. The urgency stems from the uncontrolled manner in which the "experiment" is being conducted. The release of fossil carbon as CO_2 has been increasing at an exponential rate since the beginning of the industrial revolution about 100 years ago (Fig. 1). As a result the concentration of CO_2 in the atmosphere, which thus far has grown only about 12%, may double in the next 60 years or so. The effects may well become visible suddenly and, because of the great momentum developed by the machinery that produces man's energy, could grow out of control before remedial actions become effective.

The principal effect of an increased concentration of CO_2 in the atmosphere should be a warming (Schneider 1975). While CO_2 is transparent to the incoming solar radiation, it absorbs a portion of the infrared radiation returned to space by the Earth. This "greenhouse effect" has given the words themselves a rather specialized meaning. The amount of warming produced by a given increase is, as we shall see, still quite uncertain, involving as it does all the complexities of the world climate. Yet the impacts of increased atomspheric CO_2 on man's environment could be large indeed, rendering this a problem in impact assessment of unprecedented scope and difficulty.

This matter has been considered extensively in the current literature by climatologists (e.g. Mitchell 1972, 1975; Schneider 1975), geochemists (e.g. Broecker 1975; Keeling, in press), and biologists (e.g. Reiners et al.

1973), but not often in all its important aspects. In the present article we shall attempt to consider these and especially the important uncertainties that presently limit the reliability of impact assessment. Effective remedial actions, if and when they become necessary, quite obviously cannot be expected unless and until their need can be foretold with a reliability that will match the considerable costs of their implementation.

The growth of atmospheric CO_2

Since the beginning of accurate and regular measurements in 1958, the concentration of CO_2 in the atmosphere has shown an accelerating increase upon which is superimposed annual fluctuations from photosynthesis and other seasonal effects (Fig. 2). The current average value is about 330 ppm, compared to estimated preindustrial values between 290 and 300 ppm (Callendar 1958). The measurements in Figure 2 were taken at Mauna Loa Observatory (3,400 m elevation) in Hawaii and were corrected for any temporary disturbances from local sources. Measurements at Point Barrow, Alaska, from aircraft over Sweden, and at the South Pole all show quite clearly the same secular increase (Machta et al. 1976).

The excess of the annual average concentration over the preindustrial value has grown about 4% per year (lower curve in Fig. 3). Over the same period the cumulative amount of CO_2 produced by the burning of fossil carbon (upper curve in Fig. 3) has been about twice as great as the atmospheric increase and has shown a similar rate of growth. (Except for brief interruptions during the two world wars and the great depression, the increase in the rate of release of fossil carbon has been 4.3% per year.) This suggests that, on the average, about half of the fossil carbon flux has been balanced by all the other fluxes in the carbon cycle, the excess carbon being stored in the oceans and the terrestrial biomass.

The correlation of the two curves in Figure 3 is the strongest evidence to date that the fossil carbon flux is primarily responsible for the secular increase in atmospheric CO_2. While perhaps it is not conclusive evidence, it certainly gives ample reason for concern. Will the fraction of released fossil carbon that, in effect, remains airborne increase, remain the same, or decrease? How accurately can we predict the future course of the CO_2 concentration in the atmosphere? Answers to these questions require a knowledge of the future rate of consumption of fossil fuel and of the response of the carbon cycle.

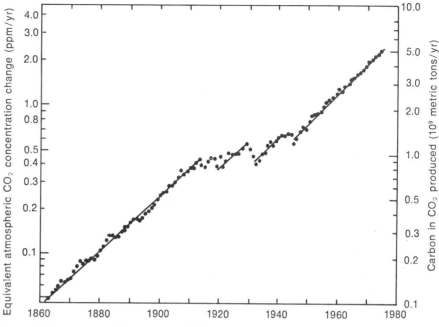

Figure 1. The annual world production of CO_2 from fossil fuels (plus a small amount from cement manufacture) is plotted since the beginning of the industrial revolution. Except for brief interruptions during the two world wars and the great depression, the release of fossil carbon has increased at a rate of 4.3% per year. (1860–1949 data from Keeling 1973a; later data from Rotty 1976.)

The carbon cycle

At the current concentration of 330 ppm of CO_2, the atmosphere contains about 700 Gt (1 Gigaton = 10^9 metric tons) of carbon (Fig. 4). This is substantially less than the carbon stored in living and dead biomass on land (about 1,800 Gt), somewhat more than that stored mostly as inorganic carbon in the well-mixed surface waters of the ocean, and much smaller than that stored in the deep oceans (about 32,000 Gt). The fluxes of carbon between the land and the atmosphere from photosynthesis (gross primary production, see Table 1) in one direction and respiration, decay, and fires in the other are estimated to be about 113 Gt/yr, and between the oceans and atmosphere, about 90 Gt/yr. Thus substantial portions of the carbon in the atmosphere, in the surface waters of the oceans, and on land are circulated each year in the carbon cycle. Quite obviously the relatively small amount in the atmosphere can be appreciably influenced by any changes in the major fluxes and pools of the cycle.

Most of the land biomass is present as relatively slowly exchanging material: humus and recent peat (about 1,000 Gt) and larger, long-lived stems and roots of vegetation (about 600 Gt). Only a relatively small fraction (about 160 Gt) is present as rapidly ex-

changing material: small stems and roots, litter, etc. These pools of slowly and rapidly exchanging carbon are allocated to groups of ecosystems in Table 1 and Figure 5. Also shown in Table 1 are the fluxes corresponding to gross photosynthesis (GPP) and, after subtraction of green plant respiration (R_g), the net production (NPP). Included as well are our estimates of the fluxes from fires (F) and from heterotrophic respiration (R_h) of animals and decomposers. (Heterotrophs are organisms that do not fix CO_2.)

It is clear from Table 1 and Figure 5 that man can have a significant influence on the fluxes between the land and the atmosphere. If, for example, he could cause the living biomass (about 600 Gt) to increase at a rate of 1% per year, this would more than counterbalance the current annual production of CO_2 from fossil fuel (5 Gt/yr). Since woods have more carbon per hectare, this could be accomplished by conversion of more land to woods. However, the maximum increase in biomass that could be realized would be small compared to the total mass of fossil carbon (perhaps 7,300 Gt) that man might ultimately consume.

Actually it is more likely that the biomass is being reduced by the ac-

tivities of man, particularly in the southern woods (south of 30° N latitude) where the traditional cycle of slash-and-burn agriculture may be shortened because of the pressures of growing population to the point where insufficient time is allowed for forest regrowth and soil replenishment during the fallow part of the cycle. As a result, a net conversion of woods to nonwoods may be taking place there. It is quite possible that the conversion rate is as much as 1% a year. Because of the lower concentration of carbon in nonwoods (Fig. 5) this could amount to a net flux of more than 1 Gt/yr of CO_2 to the atmosphere.

Another effect to be considered is the enhanced rate of photosynthetic production that might be caused by the increasing concentration of CO_2 in the atmosphere. Controlled studies of plant growth show that there is such an effect when other nutrients are not limiting (Allen et al. 1971); however, its importance in the carbon cycle is presently unclear. The common practice in modeling has been to assume that the fractional increase in the rate of photosynthesis is equal to the fractional increase in CO_2 times a factor β, which is less than unity as a result of deficiencies in other nutrients or water. Keeling's (in press) models seem most consistent with β

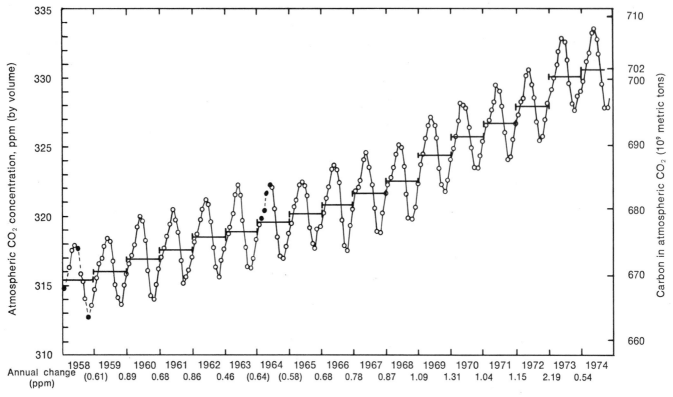

Figure 2. Monthly average values of the concentration of CO_2 in the atmosphere at Mauna Loa Observatory, Hawaii, are plotted since the beginning of accurate and regular measurements in 1958. A few of the annual changes are enclosed in parentheses because they include interpolated points, shown as dots. Variations in photosynthesis and other seasonal effects produce the annual cycle. Mean annual concentrations are well above the preindustrial level (290–300 ppm), and the secular increase is quite apparent. (1958–1971 data from Keeling et al. 1976; later data by pers. comm. from Keeling.)

β = 0.27, while Oeschger et al. (1975) found values of β from 0 to 0.4 consistent with their model. This range corresponds to a 0 to 2.7 Gt/yr increase in the biomass carbon resulting from the 12% increase in atmospheric CO_2 that has occurred over the last 100 years.

The principal forms of carbon in the ocean are inorganic; i.e. "carbonic acid" (dissolved CO_2 plus H_2CO_3), bicarbonate ion (HCO_3^-) and carbonate ion ($CO_3^=$), amounting to about 1%, 89%, and 10%, respectively, of a total concentration of about 0.002 moles/liter and a total amount of about 39,000 Gt of carbon. The next most abundant dispersed form is dead organic matter, highly variable in concentration, averaging roughly 10^{-4} moles of carbon per liter and totaling about 1,650 Gt of carbon. It appears to exist in a continuum of particle sizes down to small organic molecules (Wangersky 1972). The living organic matter, principally plankton and confined largely to the surface region, is quite minor in amount (about 1 Gt of carbon, an estimate between those of Strickland [1965] and Whittaker [1975] or Bolin [1970, 1974]). Another reactive form of carbon is the carbonate solids (mostly $CaCO_3$ formed by sea organisms), both suspended and in the surface layer of the bottom sediments. The top 10 cm or so of these deposits, amounting to 400 Gt of carbon (W. Broecker, pers. comm.), is mixed sufficiently by the foraging activities of bottom animals so that it may reasonably be included as a reactive part of the ocean reservoir.

Only the surface layer of the oceans, extending to a depth averaging about 70 m, is heated by the sun and agitated by the wind such that it is relatively well mixed. Beneath this is a stagnant region, the thermocline, stabilized by a decreasing temperature and an increasing density to a depth of about 1,000 m. Below this is the much larger region of the cold (<5°C) deep ocean, isolated from the surface waters by the thermocline. However, when the surface waters are sufficiently cold and/or saline the stabilizing density gradient weakens and the surface waters can sink (or mix) to various depths and spread horizontally. This produces a worldwide circulation that involves descending surface waters mostly in polar regions and upwelling of deep water elsewhere.

The capacity of the surface waters alone to take up atmospheric CO_2 is determined largely by the reaction of CO_2 with carbonate ion to form bicarbonate ion

$$CO_2(g) + CO_3^= + H_2O \rightleftarrows 2HCO_3^-$$

As for the other reactions, the amount of neutral "carbonic acid" that can form is small, and the amount of CO_2 converted to HCO_3^- is limited by the constraint that the concentration of other ions must change to preserve the balance of ion charges.

To explore the consequences of this situation more fully, let us assume as an approximation that the above reaction is the only one occurring and that $CO_3^=$ and HCO_3^- are the only forms of carbon present in the ocean. The equilibrium condition for the reaction is

$$Q = \frac{[HCO_3^-]^2}{P_{CO_2}[CO_3^=]}$$

where Q is a constant at a given temperature. (Q is also affected slightly by small variations in the salinity of sea water.) Appropriate differentiation of this expression (with ion charge and material balances preserved) tells us that the ratio of fractional changes in the atmospheric pressure (P_{CO_2}) and the total dissolved carbon ($\Sigma C = [HCO_3^-] + [CO_3^=]$) is given solely by the ratio $R = [CO_3^=]/\Sigma C$.

$$\frac{d \ln P_{CO_2}}{d \ln \Sigma C} = \frac{1 + 3R}{R(1 - R)}$$

Since R is fairly small ($\sim$0.13), this ratio of fractional changes (called the buffer factor) is fairly large ($\sim$12.3). A more accurate calculation (described by Keeling 1973b), including contributions from other forms of carbon, gives a lower buffer factor, near 10. But the important point is that the capacity of the surface waters alone to take up CO_2 is quite limited, determined primarily by the small supply of $CO_3^=$ ion present. Since the surface waters contain an amount of carbon comparable to that in the atmosphere, less than a tenth of the current fossil carbon flux could be taken up by the surface waters alone, and this fraction should decrease as the carbonate ion is consumed.

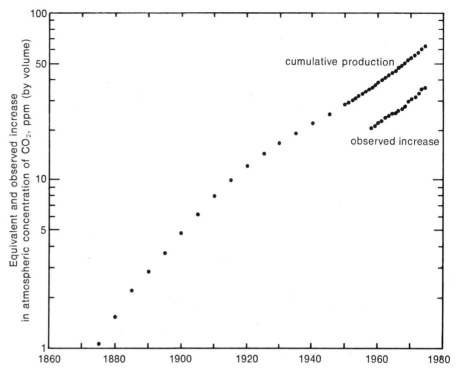

Figure 3. The cumulative production of CO_2 since 1860 (i.e. the summed changes from Fig. 1 expressed as the equivalent atmospheric concentration) is compared with the observed increase in the mean annual concentration (from Fig. 2) since that time. The similarity in the rates of increase (about 4% per year) provides strong evidence that these two quantities are related. About 50% of the fossil carbon flux apparently has been balanced, at least since 1958, by a flow of CO_2 to the oceans and/or the land biota. (The 1860 atmospheric concentration was assumed to be 295 ppm.)

Clearly if the oceans are a sink for a substantial part of the fossil carbon flux, this must be caused by the mixing that occurs between the surface waters and the deep waters. The distribution in the oceans of ^{14}C, which originates in the upper atmosphere from bombardment of ^{14}N by cosmic rays or from nuclear tests, indicates that the average residence time of water in the deep oceans is in the range of 500 to 2,000 years (Broecker and Li 1970; Oeschger et al. 1975). This is equivalent to circulation of only 2% to 8% of the surface water per year to the deep ocean. Hence simple box models, in which each region is assumed to be homogeneous, should not predict much capacity of the oceans to respond to the relatively rapid increases in atmospheric CO_2. Oeschger et al. (1975) point out, however, that if the mixing of surface water with deep water is assumed to occur by eddy diffusion, then the increase of CO_2 in the atmosphere can establish a concentration gradient of carbon to the deep oceans—a gradient that can store large amounts of CO_2, sustain a higher flux, and perhaps account for the entire amount that has left the atmosphere, equiv-

alent to about 50% of the fossil carbon flux.

The ultimate capacity of the ocean system, including the $CaCO_3$, is far in excess of that required to deal with all the fossil carbon that mankind may wish to use. Indeed, as Sillén (1963) and Holland (1972) have pointed out, in the long run the concentration of CO_2 in the atmosphere should tend to remain near the present level by virtue of the reaction

$$CaCO_3(c) + CO_2(g) + H_2O$$
$$\rightleftarrows Ca^{2+} + 2HCO_3^-$$

and the equilibrium condition

$$Q = [Ca^{2+}][HCO_3^-]^2/P_{CO_2}$$

since the amounts of Ca^{2+} and HCO_3^- ion and $CaCO_3$ solid in all the oceans is far greater than the amount of fossil carbon. However, this natural control mechanism may be far too sluggish to cope with the high rate of fossil-fuel use.

Future levels of CO_2

The future growth of CO_2 in the atmosphere should depend primarily on the rate of fossil fuel consumption

and on the manner in which the carbon cycle responds to the resulting flux of CO_2. In view of the large uncertainties involved, we will consider here only high and low scenarios for fossil fuel use and ranges of response of the two largest fluxes of the carbon cycle, i.e. between the atmosphere and the oceans and between the atmosphere and the land.

Current estimates of recoverable fossil fuel suggest that about 5,600 Gt of carbon reside in forms other than oil shale (Averitt 1975; Hubbert 1974). The amount of carbon from shale containing more than 25 gallons of oil per short ton (104 liters per metric ton) is perhaps 1,700 Gt (Duncan and Swanson 1965), giving a total economic reserve of about 7,300 Gt of fossil carbon. If we were to assume, quite unrealistically, that the use of fossil fuel will continue to grow 4.3% each year until the supply is exhausted, then the Fossil Carbon Age would last 97 more years and the use rate at the very end would be over 300 Gt/yr, almost 64 times the current rate. The total CO_2 injected into the atmosphere in less than a century would be about twelve times the

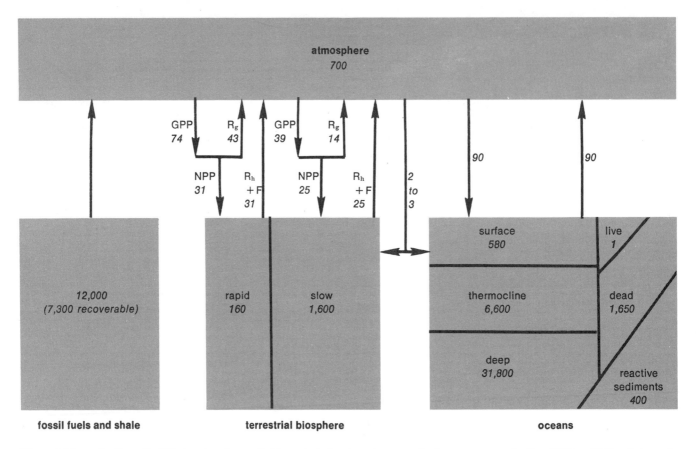

Figure 4. The major fluxes (in Gt/yr) and pool sizes (in Gt) of carbon are shown for the carbon cycle. Fluxes include gross primary production (GPP), green plant respiration (R_g), net primary production (NPP = GPP − R_g), respiration by heterotrophs (R_h), and fires (F).

preindustrial content of the atmosphere. Near the end the amount added every two years would equal the carbon content of the preindustrial atmosphere.

A much more reasonable assumption for the high-use scenario is that an initial growth rate of 4.3% per year will be reduced in proportion to the fraction of the ultimate supply of fossil fuel that has been used. This scenario (upper curve in Fig. 6) is equivalent to one of Keeling's (in press) projections. Since it predicts that more than half of all the fossil carbon will be released in less than 100 years, it still represents a strong perturbation of the carbon cycle.

For the low-use scenario let us assume that the growth rate will be only 2% per year until 2025, to be followed by a symmetrical decrease as renewable energy sources become more available and the use of fossil fuel is discouraged (lower curve in Fig. 6). The total fossil carbon released in this scenario is about one quarter that of the high scenario and about 1.5 times the carbon content of the preindustrial atmosphere.

But how will the carbon cycle respond to the flux of fossil carbon? If we make the assumption that it will continue to respond as in the recent past by taking up 50% in the oceans and/or the land biota, then we obtain the curves in Figure 7 for the growth of CO_2 in the atmosphere. This assumption may at first sight seem unduly pessimistic since the carbon cycle might well respond more strongly to prevent such large excursions above the "normal" atmosphere concentration of CO_2. Yet man's conversion of the forests to agricultural use as well as the decreasing capacity of the surface waters of the oceans to take up CO_2 could outweigh this tendency.

Without a more certain knowledge of how the other fluxes of the carbon cycle achieve the present net removal rate, perhaps the best that can be done is again to take two extreme cases and estimate the variations possible in the largest fluxes and the consequent effects on future levels of atmospheric CO_2. In the first case, population pressure forces the conversion of all arable land to food production in 100 years, thus con-

verting a substantial fraction of the standing crop of wood and a smaller fraction of the humus to CO_2. This could amount to as much as 200 Gt of extra carbon to the atmosphere, and over the next century the net flux would average 2 Gt/yr. In the second case, with decreasing growth of population, more efficient agriculture, increasing use of biomass as fuel (with planned regrowth), and less wasteful burning, a flux averaging 2 Gt/yr in the opposite direction might be achieved—especially if photosynthetic productivity increases with greater atmospheric CO_2 level.

On the basis of these two cases we assign a range of ±2 Gt/yr to plausible changes in the net flux between the land and the atmosphere averaged over the next 100 years. The resulting effect on the cumulative anthropogenic release of CO_2 and on future levels of CO_2 in the atmosphere (Fig. 7) is appreciable but fairly small compared to the possible variation from the fossil carbon flux.

The oceans are thought by many to be the sink that accounts for the largest part of the 50% equivalent of the anthropogenic flux that leaves the atmosphere. Because the capacity of the surface waters to take up CO_2 is so small, such an uptake would seem to require the supply of considerable additional capacity from the deep oceans. The net consumption by photosynthesis seems also to depend on the supply of certain limiting nutrients (e.g. phosphorous and silicon) from the deep oceans (Broecker 1974). Present knowledge of ocean circulation is insufficient to predict with any certainty how the oceans will behave as a sink for CO_2. We assume in Figure 7 that the uncertainty in net flux to the oceans will be equivalent to ±10% of the fossil carbon flux.

Climate change

A long history of natural climate change is revealed by the geologic record. In a recent review (ICAS 1974), fluctuations of the average surface temperature over the past one million years were represented approximately as the sum of sinusoidal variations with different periodicities (Table 2). The two longest periods may be caused by similar periodicities in the fluctuations of the shape of the Earth's orbit (Mitchell 1972). Attempts to correlate climate variation with other natural causes, such as sunspot cycles and related variations in the solar constant (Schneider and Mass 1975) seem less well founded.

We are presently in a very warm period, about 10,000 years since the end of the last glacial. In the last 100 years the global mean temperature rose about 0.6°K to 1940 and fell about 0.3°K in the following three decades (Mitchell 1961, 1972). The important point from the previous history of climate change is that natural fluctuations which occur rapidly enough (i.e. over a few generations) to produce easily recognized impacts on the affairs of man are likely to be of small amplitude. Over the next 100 years the mean global temperature may be expected to change no more than 0.5° to 1.0°K from natural causes. Broecker's (1975) analysis suggests that after 1980 there will be an onset of natural warming (if warming does not occur sooner because of increased atmospheric CO_2).

Any activity of man that can alter significantly the radiation balance of the Earth can affect the climate. In addition to the effect of increased

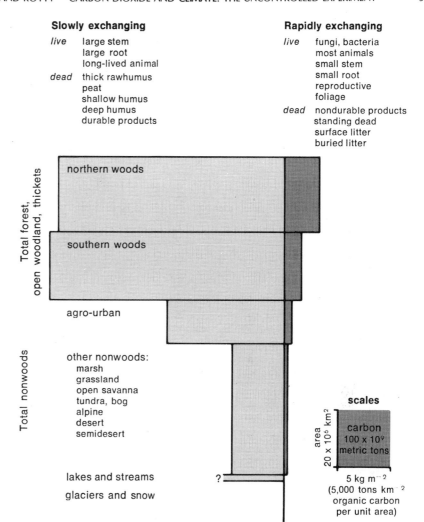

Figure 5. The quantities of carbon in the various terrestrial ecosystems are shown as concentration (vertical direction) vs. area (horizontal direction). The vertical line divides rapidly exchanging from slowly exchanging organic forms.

CO_2, there is the growing direct production of heat that results from man's use of energy, but this is presently an insignificant 0.01% of the solar flux and important only in areas of high energy utilization. There is the local effect of paved areas that alter the reflectivity (albedo) of the Earth's surface. A possibly important anthropogenic effect is that of airborne particulates which, it has often been suggested, might produce a cooling by backscattering solar radiation. But actually the magnitude of this effect, even whether it is one of cooling or warming, depends on many factors (Charlson and Pilat 1969). These include the altitude, size, concentration, and reflectivity or refractive index of the particles. Because of this the impact of anthropogenic particulates is quite uncertain. Those produced by the consumption of fossil fuels are relatively large, remain at low altitude, and are airborne but a short time. As a result, their effect on cli-

mate may remain a local one long after the effects of CO_2 become quite apparent.

The impact on climate to be expected from a particular stimulus has been deduced from models developed over the last several decades, from simple one-dimensional radiation-balance models of the Earth's atmosphere to quite complicated so-called general circulation models that include the circulation of the atmosphere in three dimensions. Such models indicate an increase in temperature with increased CO_2 and suggest that the temperature response is approximately logarithmic, i.e. each doubling in the concentration of CO_2 produces about the same increase in the average temperature of the troposphere. The more sophisticated models predict, moreover, that the temperature change is much greater at high latitudes. The amount of the average increase in temperature per doubling

Table 1. Terrestrial organic carbon pools and their fluxes[a]

Major global pool[b]	Turnover rate	Carbon inventory[c] (Gt)	Recent fluxes (Gt/yr)					
			Photosynthesis			Respiration		Fire[f] Fe
			NPP[d]	GPP/NPP	GPP[d]	R_g[d]	R_h[e]	
Northern woods (N of lat. 30° N)	slow[c]	540	8			(3)	(7)	(1)
	rapid[c]	70	10			(6)	(9)	(1)
		610	18	1.5?	(27)	9	16	2
Nonwoods (world)	slow	540	8			(3)	(8)	(0)
	rapid	40	10			(15)	(8)	(2)
		580	18	2.0?	(36)	18	16	2
Southern woods (S of lat. 30° N)	slow	520	9			(8)	(7)	(2)
	rapid	50	11			(22)	(10)	(1)
		570	20	2.5?	(50)	30	17	3
Subtotals	slow	1,600	25			14	22	3
	rapid	160	31			43	27	4
		1,760	56		(113)	57	49	7

[a] Excluding incipient fossil fuel, buried humus, etc. (carbon with decay rate < 0.001 yr^{-1}). Estimates and geographic breakdown are slightly modified from Olson 1970; SCEP 1970, pp. 160–66; Reiners et al. 1973; Bazilevich 1974; and Lieth and Whittaker 1975.

[b] Woods (i.e. forest, open woodland, and woody swamp) is partitioned because it is such a large pool and because the seasonal contrast of photosynthesis and respiration on lands north of 30° N latitude is more striking than in south temperate, subtropical, and tropical zones.

[c] Rapidly exchanging pools (turnover averaging a few years) are estimated significantly higher than SCEP (1970, Table 2.A.2); slowly exchanging pools (many turnover times, averaging several decades) are only slightly higher than SCEP.

[d] The net primary production (NPP) estimate of SCEP is provisionally retained, but this and especially the allocation to pools and the indirect allowance for green plant respiration (R_g) and hence gross primary production (GPP) or photosynthesis are very tentative. Many literature estimates of photosynthesis are low because they deliberately or inadvertently omit the rapidly exchanging pools, overlook many processes of rapid turnover, or use old or conservative estimates of production.

[e] Heterotrophic respiration (R_h) of animals and decomposers (bacteria and fungi) recycles to atmospheric CO_2 that part of NPP which is not first burned (F), exported, or stored (e.g. in pools with decay rate < 0.001 yr^{-1} or mean residence time $> 1,000$ years). Such storage and export to oceans are assumed $< 1 \times 10^9$ tons/yr.

[f] The carbon fluxes from fires, which are very uncertain, were estimated assuming that in each woods pool a period averaging about 200 years elapsed between consuming fires in a given location.

of CO_2 (ΔT) given by the various models (Schneider 1975) are in the range 0.7 to 9.6°C. Schneider (1975) critically reviewed these and concluded that "a state-of-the-art, order-of-magnitude estimate" is in the range 1.5 to 3°K. He notes, however, that this estimate could prove to be high or low by severalfold as a result of "climate feedback mechanisms not properly accounted for in the state-of-the-art models."

Some mechanisms not adequately accounted for in present models include (1) *decreased snow and ice coverage*—a positive (amplifying) feedback since the resulting decrease in reflected radiation will produce further warming; (2) *changes in cloud cover and in the temperature of cloud tops*—thought to be the most important feedback mechanism not adequately treated; (3) *ocean coupling*—which cannot be fully accomplished without a better model of

ocean circulation, which itself is driven by the latitudinal temperature gradient; and (4) *land coupling*—which should include the effects of local changes in albedo and water balance as well as the direct effects of biota change on the levels of CO_2.

In addition to their differing magnitudes, these feedbacks have widely differing response times, from perhaps a few months for significant changes in average cloud cover to perhaps hundreds of years for significant changes in the extent of the polar ice caps. Since the sum of these feedback effects as a function of time is presently unknown, we will allow for their contribution to the relatively rapid climatic response by increasing Schneider's estimate of the range of the average ΔT to 1° to 5° per doubling of the CO_2 content of the atmosphere. This range of uncertainty in ΔT is combined with the uncertainties associated with the growth in

atmospheric CO_2 to produce the range of projections in Figure 7 of the increase in the average temperature over the next 100 years.

Temperature changes near the upper limits of these projections—for a high fossil-fuel use scenario and a ΔT near 5°C per doubling of atmospheric CO_2—would in less than 100 years exceed the amplitude of the glacial-interglacial cycle normally traversed in tens of thousands of years. A climatic change of such unprecedented rapidity can hardly be viewed as anything less than catastrophic. The changes projected near the lower extremes of Figure 7 are possibly acceptable, though even here the increases in global average temperature from anthropogenic causes may become larger than natural variations in the fairly near future.

Thus mankind confronts an increasingly familiar kind of dilemma: while the impending consequences of his activities are not so clearly unacceptable that emergency measures are called for, the probability that they might be so is great enough to demand strong action. Greatly increased efforts should be made to foresee more clearly the effects of anthropogenic CO_2.

Impact of climate change

Let us consider briefly some of the ways in which a general climate warming could affect man and his environment. These will obviously depend primarily upon the nature, the magnitude, and the rate of change of regional climates. Since at present such changes are even less predictable than the average warming, none of the effects can be estimated quantitatively and many may not even be foreseeable. We list here some that seem plausible.

At first thought a warmer climate might seem to be generally beneficial, with longer growing seasons that produce more food and milder winters that help save fuel. However, any rapid change in a regional climate is more likely to produce detrimental effects that far outweigh the beneficial ones. This follows simply because crops and other species have become tuned by experience and selection to match existing conditions. Most rapid changes will produce dislocations that reduce biological fitness and/or productivity before human or natural readjustments can become effective.

Dislocations from warming can take many forms. Rates of respiration and decay may increase faster than photosynthetic production. Microbial and insect pests from adjacent regions or endemic pests previously held in check may be favored. Since the average rate of evaporation should increase, a surface drying may occur in all climate zones but the wettest.

Year-to-year fluctuations in local weather is a fact of life in agriculture. Longer-term fluctuations have produced considerable economic dislocation and human suffering in developed countries and massive starvation in less developed regions. These larger, evidently natural, fluctuations are likely to produce ever more serious effects as the growing world population presses agriculture

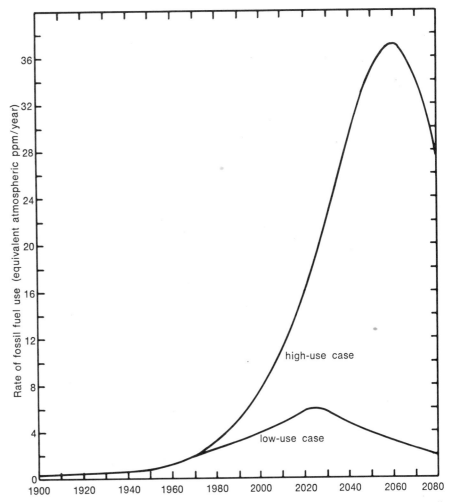

Figure 6. The two scenarios shown for the use of fossil fuels over the next 100 years are suggested as plausible limiting cases. With the high-use case, assuming an annual growth rate of 4.3% reduced in proportion to the fraction of the ultimate fossil fuel supply that has been used, more than half of all fossil fuel carbon will be released in less than a century. With the low-use case, assuming a 2% growth rate till 2025 followed by a symmetrical decrease, only about a fourth as much fossil carbon will be released as in the high-use case.

to its limits of productivity. Clearly, then, still larger unfavorable changes in some regional climates are likely to produce dire consequences indeed, and the more so if they appear suddenly and are not anticipated.

Perhaps the most often cited effect of a general warming is the rise in sea level that would accompany the melting of glacial ice. A complete melting could raise the level of the oceans over 50 m, and quite obviously even a partial melting could have profound effects on the shorelands of the world and all their associated

Table 2. Natural climate fluctuations represented as the sum of cyclic changes

Characteristic period (yrs)	Fluctuation in average temperature (°K)	Last temperature maximum (yrs ago)	Maximum rate of temperature change (°K/yr)
100,000	8.0	10,000	±0.00025
20,000	3.0	8,000	±0.00045
2,500	2.0	1,750	±0.0025
~200	0.5	75	±0.0075
~100	0.5	35	±0.015

SOURCE: ICAS 1974

values. The melting process, however, is quite complex and not yet understood well enough to predict the rate at which this could occur. Warmer air alone could produce only a slow melting (G. Marland, pers. comm.), but warm polar waters could induce a flow of ice from the continental shelves into the sea, possibly raising the sea level 5 m in 300 years (Hughes, cited by Gribbin 1976).

In contrast to a slowly rising sea level, inland lake levels and water supplies may be lowered more rapidly where climate is drier. For example, navigation, sewage, and water problems already experienced as a result of the low level of the Great Lakes in 1964 could become more costly in the future.

Diminished ocean circulation and weakened upwelling and nutrient replenishment around some fertile fisheries might add to the stress they already experience from overfishing. Other marine and freshwater food chains could become even more vulnerable because of impacts on nursery and feeding areas of the changing shores, rivers, and lakes.

These and other effects will have sociological and political repercussions that are difficult or impossible to predict and evaluate fully. Because costs and benefits of climatic change will fall very unevenly on different regions, groups perceiving themselves as unthreatened or even as beneficiaries of the change will question the validity or cost of steps that might be recognized elsewhere as desirable for correcting the situation. Groups quickly perceiving their own losses, and fearing more drastic changes yet to come, could look for scapegoats long before the several links from energy to atmospheric CO_2 to climate to ecological and social impact can be regarded as settled from a scientific standpoint. Many of these impacts will affect billions of people to a significant degree, and almost any summation of the possible social cost from excess CO_2 becomes far too great to languish in the realm of academic debate. The possibility that many may find some benefits in the total change in no way dismisses the imperative to estimate and balance the costs and benefits. Finally, there is also the serious ethical question of the commitment that we and our children may be making for scores of future generations to changes in the world environment that seem impossible to reverse for many centuries to come.

What is to be done?

As Revelle and Suess (1957) have said, "Man may learn much about the processes that determine weather and climate by observing carefully the course of the CO_2 experiment." But, of course, this in no way diminishes the present urgency. Man is not conducting a controlled experiment; he is using cheap, abundant fuel and will continue to do so until the impending consequences clearly become more costly than alternative sources of energy. Hence the most urgent thing to be done is to improve our ability to predict these consequences. The problem of how to do this seems to resolve itself into three parts.

1. *We must learn more about the carbon cycle.* In particular, we must learn to predict the fluxes of CO_2 between the atmosphere and the oceans and between the atmosphere and the land. This will require a greatly increased effort to monitor, study, and model the processes of the ocean and the land that control these fluxes.

For the ocean, the most important process is the circulation and mixing of surface water with the deep water. A greatly increased ocean monitoring effort seems to be called for, one that would provide information more comparable in detail to that of the atmosphere. It should include the continued study of the distribution of tracers such as ^{14}C and ^{3}H and the application of whatever other techniques that can give information on deep circulation and also on $CaCO_3$ reaction rates. In all ocean monitoring activities, the collection of data and the development of circulation models should be closely coupled.

For the land, the processes that most strongly influence the net flux of CO_2 to or from the atmosphere, aside from the burning of fossil fuel, are those of photosynthesis, respiration, decay, and the burning of biomass. Here a monitoring and modeling effort at least as extensive as that for the oceans may be needed. Of greatest importance is the effect of increased atmospheric CO_2 on photosynthetic production rates, the potential effects from increased land use for agricul-

ture, and the turnover rates of live and dead organic matter. The overall objective should be to predict accurately the change in the net flux of CO_2 to the land for a given scenario of fossil fuel use and land use.

2. *We must predict more accurately the climatic effects of increasing atmospheric CO_2.* Quite clearly this is the major uncertainty in the assessment of the environmental impact of CO_2. Perhaps the only means that will provide the needed predictive power is a reliable model, one that can reproduce the important features of each regional climate when the driving variables of world climate are properly chosen.

Climate models have developed rapidly over the last two decades, and the most elaborate have reached the limits of capacity and speed of the largest and fastest computers available. Advances in computer technology now in sight may permit reductions in running time of perhaps a factor of ten, and Mitchell predicts (pers. comm.) that much more detailed global climate models will then be developed. It is his hope that insights thus gained will lead to simplified "smart" models that can make the needed prediction more elegantly and accurately. Many feel that improved models must include a better representation of the basic physical processes of climate, e.g. fluid dynamics, cloud formation, and cloud interactions. Such may come from an improved basic understanding, or perhaps these processes can be represented adequately by better empirical representations.

To support these development efforts and also to validate properly a reliable climate model, a more extensive observational network—including satellite monitoring as well as surface-based measurement—will be needed that provides more detailed and accurate worldwide meteorological data. The need for climate prediction is so great and the difficulty of achieving it so formidable that a much larger climate study effort should be undertaken; a broad program that includes all aspects of climatology. It should certainly include numerical modeling, physical modeling, basic studies of fluid dynamics, studies of cloud formation and interactions, systematic surface observations of climatic variables throughout the world, atmospheric monitoring by satellites, and statistical analysis of observational data. The staff for such a program should include, in addition to meteorologists and climatologists, representatives from the physical and engineering sciences. What seems most needed (S. Manabe, pers. comm.) for rapid progress in climate research are new ideas and insights that might replace the "brute force" methods of numerical computation with elegant simplifications—ideas and insights that are most likely to come from the bringing together of workers with diverse backgrounds.

3. *We must anticipate the consequences of climate change.* Ecological, economic, social, and political impacts must be foreseen quickly and accurately enough to guide remedial actions. The sooner specific climate changes and their effects can be foreseen, the more likely it is that effective actions can be taken. Assessments and supporting research of such a highly interdisciplinary nature and scope will probably require more effective information handling and modeling plus methods for validating models and scenarios that will enhance credibility.

In short, much of the problem is to develop and then test and use the *means* for credible assessment of impacts. Biophysical ecology (Gates and Schmerl 1975) and physiology provide valid principles for estimating how organisms can adapt. Models for species populations should be modified for crops and natural communities so that shifts in life zones and rates of production can be developed for test regions and eventually larger areas of continents and oceans. Beyond demonstrating that models are consistent with data from which they are formulated and tuned, testing their consistency with independent experience is needed for suggesting the kinds of perturbations to be expected from a known or hypothetical climate change.

If the present predictions (Fig. 7) are correct, not long after the year 2000 the warming effect of increased atmospheric CO_2 could become conspicuous above the "noise level" from other causes of climate fluctuation. However, the momentum of societal fuel-use patterns may make it difficult then to adjust from fossil energy

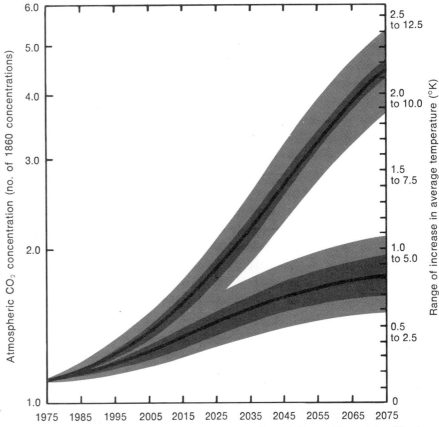

Figure 7. The projected growth in atmospheric CO_2 and the possible increase in the average surface temperature of the Earth are shown for the next 100 years. The solid curves represent the limiting scenarios in Figure 6 with 50% of the fossil-fuel flux balanced by ocean and/or land uptake of CO_2. The outer bands represent the effect of 40% to 60% uptake. The inner bands represent the effect of a ± 2 Gt/yr variation in the net flux to the land. The range of the increase in average temperature corresponds to 1 to 5° K per doubling of the CO_2 concentration.

to nonfossil energy quickly enough to avoid eventual severe consequences. Hence the time available for action may be quite limited.

Quite clearly we must improve our predictions of the consequences of increased atmospheric CO_2. The largest uncertainty is the specific effect of a given increase on the regional climates of the world. As a consequence, a first priority should be given to the study of possible climate changes. A better understanding of the carbon cycle is also needed to project better the rate of increase of atmospheric CO_2. Finally, we must learn to project the impact of the climatic changes on man, his environment and his society.

As the potential consequences of various scenarios of energy development become more clearly foreseeable, these must be included in the cost-benefit analyses. Depending upon the severity of our energy problems, it may be expected that

nonfossil fuel options, such as fusion and breeder reactors for central station generation of electric power and wind, solar, and geothermal energy for dispersed sources, will be increasingly brought into use. One nonfossil fuel option being more actively considered is the use of cultivated and waste biomass as fuel, perhaps with conversion to methanol, ethanol, methane, and/or hydrogen (Pollard 1976). This could become an attractive "solar energy" conversion method that recycles atmospheric CO_2.

While actions that reduce the impact of climate change such as the establishment of food reserves and the diversification of agriculture could be effective even if taken unilaterally by individual nations, other actions that reduce the rate of production of anthropogenic CO_2 depend strongly on multinational cooperation to be effective. If the severe economic and political repercussions that are likely on a world scale are to be avoided, a technological commitment must be made in the next few years and a world strategy arrived at with enlightenment and wisdom. Though humanity may not be able to foresee the consequences of "the great experiment" clearly enough to control them, we cannot afford not to try!

References

Allen, L. H., Jr., S. E. Jensen, and E. R. Lemon. 1971. Plant response to carbon dioxide enrichment under field conditions: A simulation. *Science* 173:256.

Averitt, P. 1975. World estimated total reserves of coal. *USGS Bull.* 1412.

Bazilevich, N. I. 1974. Energy flow and biogeochemical regularities of the main world ecosystems. In *Proc. 1st Intl. Cong. of Ecol.*, p. 182. Wageningen, Netherlands: Pudoc.

Bolin, B. 1970. The carbon cycle. *Sci. Am.*, September, p. 125.

Bolin, B. 1974. Modeling the climate and its variations. *Ambio* 3:180.

Botkin, D. B., J. F. Jonak, and J. R. Wallis. 1973. Estimating the effects of carbon fertilization on forest composition by ecosystem simulation. In *Carbon in the Biosphere*, ed. G. M. Woodwell and E. V. Pecan, p. 328. USAEC Conf. 720510.

Broecker, W. S. 1974. *Chemical Oceanography*, p. 25. Harcourt Brace Jovanovich.

Broecker, W. S. 1975. Climate change: Are we on the brink of a pronounced global warming? *Science* 189:460.

Broecker, W. S., and Y-H. Li. 1970. Interchange of water between the major oceans. *J. of Geophys. Res.* 75:3545.

Callendar, G. S. 1958. On the amount of carbon dioxide in the atmosphere. *Tellus* 10:243.

Charlson, R. J., and M. J. Pilat. 1969. Climate: The influence of aerosols. *J. of Appl. Meteorol.* 8:1001.

Duncan, D. C., and V. E. Swanson. 1965. Organic-rich shale of the United States and world land areas. *USGS Circ.* 523.

Gates, D., and R. Schmerl. 1975. *Biophysical Ecology.* Springer Verlag.

Gribbin, J. 1976. Antarctica leads the ice ages. *New Sci.* 69:695.

Holland, H. D. 1972. The geologic history of sea water—An attempt to solve the problem. *Geochem. et Cosmochem. Acta* 36:637.

Hubbert, M. K. 1974. Estimate of the ultimate world production of petroleum liquids. Senate Com. on Int. and Ins. Affairs, Committee Print Ser. No. 93-40 (92-75).

ICAS. 1974. Report of the ad hoc panel on the present interglacial. NSF: Federal Council for Science and Technology Interdepartmental Committee for Atmosphere Science, ICAS 18b-FY75.

Keeling, C. D. 1973a. Industrial production of carbon dioxide from fossil fuels and limestone. *Tellus* 25:174.

Keeling, C. D. 1973b. The carbon dioxide cycle: Reservoir models to depict the exchange of atmospheric carbon dioxide with the oceans and land plants. In *Chemistry of the Lower Atmosphere*, ed. S. I. Rasool, p. 251. Plenum.

Keeling, C. D. In press. Impact of industrial gases on climate. In *Energy and Climate: Outer Limits to Growth.* Natl. Acad. of Sci.: Geophys. Res. Bd., Geophys. Study Com.

Keeling, C. D., R. B. Bacastow, A. E. Bainbridge, C. A. Ekdahl, Jr., P. R. Guenther, L. S. Waterman, and J. F. S. Chin. 1976. Atmospheric carbon dioxide variations at Mauna Loa Observatory, Hawaii. *Tellus* 28:538.

Lieth, H., and R. H. Whittaker. 1975. Primary productivity of the biosphere. *Ecol. St.* 14. Springer Verlag.

Machta, L., K. Hanson, and C. D. Keeling. 1976. Atmospheric carbon dioxide and some interpretation. Office of Naval Res. Conf. on the Fate of the Fossil Fuel Carbonates, Honolulu, Hawaii, January 19–23.

Mitchell, J. M., Jr. 1961. Recent secular changes of global temperature. *Ann. of the NY Acad. of Sci.* 95:235.

Mitchell, J. M., Jr. 1972. The natural breakdown of the present interglacial and its possible intervention by human activities. *Quat. Res.* 2:436.

Mitchell, J. M., Jr. 1975. A reassessment of atmospheric pollution as a cause of long-term changes of global temperature. In *The Changing Global Environment*, ed. S. F. Singer, p. 149. Boston: D. Reidel.

Oeschger, H., U. Siegenthaler, U. Schotterer, and A. Gugelmann. 1975. A box diffusion model to study the carbon dioxide exchange in nature. *Tellus* 27:168.

Olson, J. S. 1970. Carbon cycles and temperate woodlands. In *Analysis of Temperate Forest Ecosystems*, ed. D. E. Reichle. Springer Verlag.

Pollard, W. G. 1976. The long-range prospects for solar-derived fuels. *Am. Sci.* 64:509.

Reiners, W. A., L. H. Allen, Jr., R. Bacastow, D. H. Ehalt, C. S. Ekdahl, Jr., G. Likens, D. H. Livingston, J. S. Olson, and G. M. Woodwell. 1973. A summary of the world carbon cycle and recommendations for critical research. In *Carbon and the Biosphere*, ed. G. M. Woodwell and E. V. Pecan, p. 368. USAEC Conf. 720510.

Revelle, R., and H. E. Suess. 1957. Carbon dioxide exchange between the atmosphere and ocean, and the question of an increase in atmospheric CO_2 during the past decades. *Tellus* 9:18.

Rotty, R. M. 1976. Global carbon dioxide production from fossil fuels and cement, A.D. 1950–A.D. 2000. Office of Naval Res. Conf. on the Fate of Fossil Fuel Carbonates, Honolulu, Hawaii, January 19–23.

SCEP. 1970. Man's impact on the global environment: Assessment and recommendations for action. In *Study of Critical Environment Problems.* Cambridge: M.I.T. Press.

Schneider, S. H. 1975. On the carbon dioxide–climate confusion. *J. of Atm. Sci.* 32:2060.

Schneider, S. H., and C. Mass. 1975. Volcanic dust, sunspots, and temperature trends. *Science* 190:741.

Sillén, L. G. 1963. How has sea water got its present composition? *Svensk Kem. Tidskrift* 75:161.

Strickland, J. D. H. 1965. Production of organic matter in the primary stages of the marine food chain. In *Chemical Oceanography*, ed. J. P. Riley and G. Skirrow, p. 478. Academic Press.

Wangersky, P. J. 1972. The cycle of organic carbon in sea water. *Chimia* 26:559.

Whittaker, R. H. 1975. *Communities and Ecosystems*, 2nd ed. Macmillan.

"Oh, for Pete's sake, let's just get some ozone and send it back up there!"

Air Pollution Meteorology

Hans A. Panofsky

As the world's population and industrialization grow, air pollution (Figs. 1 and 2) becomes a progressively more serious problem. The control of air pollution requires the involvement of scientists from many disciplines: physics, chemistry, chemical and mechanical engineering, meteorology, economics, and politics. The amount of control necessary depends on the results of medical and biological studies.

Here we will limit ourselves only to the atmosphere's role in influencing the distribution of the materials which cause air pollution and the uses to which an understanding of meteorological factors can be put in air pollution control. Air pollution may also affect weather, but this subject will not be taken up.

The state of the atmosphere affects, first, many types of sources of pollution. For example, on a cold day more fuel is used for space heating. Also, solar radiation, which is affected by cloudiness, has an influence on smog production. Second, atmospheric conditions determine the behavior of pollutants after they leave the source or sources until they reach receptors, such as people, animals, or plants. The question to be answered is: given the meteorological conditions, and the characteristics of the source or sources, what will be the concentration of the pollutants at any distance from the sources? The inverse question also is important for some applications: given a region of polluted air, where does the pollution originate?

Finally, the effect of the pollution on the receptor may depend on atmospheric conditions. For example, on a humid day sulfur dioxide is much more corrosive than on a dry day.

Hans A. Panofsky was born in Kassel, Germany. A son of the eminent art historian and critic Erwin Panofsky, he had his early schooling in Hamburg. In the 1930's he came with his parents and brother to Princeton where his father received an appointment in the Institute for Advanced Study. He graduated A.B. in 1938 from Princeton and took his Ph.D. degree at the University of California, Berkeley, in 1942. Next he was an instructor of astronomy at Wilson College, PA, and with the Department of Meteorology, New York University (with interruptions) from 1942–51. He became Associate Professor and then Professor of Meteorology at Pennsylvania State University from 1951–66. He is now Evan Pugh Research Professor of Atmospheric Sciences there. Address: Department of Meteorology, College of Earth and Mineral Sciences, The Pennsylvania State University, University Park, PA 16802.

Here are some examples of how meteorological information can be used in connection with air pollution problems:

1. It can be used in planning locations of future sources of contaminants. At present, the planning of new industries is governed mostly by the availability of water, labor, raw materials, and transportation, but usually not by the air pollution likely to ensue. Thus, for example, industries tend to follow the bottom of river valleys, where air pollution can be especially severe. In the future, the location of new plants should be influenced also by air pollution considerations, as recommended by the Energy Policy Staff of the President's Office of Science and Technology. In the case of the nuclear industry, meteorological factors are even now considered in the planning stage.

2. According to the Clean-Air Act of 1967, over 50 air pollution control regions are to be established, in the United States, in which pollution is produced by a common group of sources. To a considerable extent, meteorological factors determine how large such regions must be and what their shapes should be.

3. If air pollution is to be reduced, it is important to know who is responsible for the pollution. For example, in a typical city 50% of the sulfur dioxide (SO_2) may be emitted by power plants, and the other 50% by individual buildings as a result of sulfur in the fuels used for space heating. But after analysis of the meteorological factors, it may turn out that only 10% of the ground-level SO_2 is due to the power plants—the rest is due to more or less local heating. The reason for such a discrepancy could be that the effective emission height of the SO_2 from large plants is quite high. Pollutants are usually fairly well diluted before reaching the ground. As a result, cutting the sulfur content in the fuel used by the plants only would have little effect on the air quality. Any decision on abatement procedures must take such characteristics into account.

4. As will be seen, during certain periods of the day and on certain days, air pollution concentration will be especially high; during such periods, emission of pollutants should be curbed drastically, either by substitution of cleaner fuels, or by reduction of operations. At present, warnings of poor conditions are prepared by ESSA (Environmental Sciences Services Administration); but heeding these warnings is voluntary in

Figure 1. Air pollution of the cities.

Figure 2. Air pollution in New York City on a bad day.

most places. Eventually, a set of controls which depend both on availability of fuel and on meteorological factors should be made compulsory.

The way in which the atmospheric characteristics affect the concentration of air pollutants after they leave the source can be divided conveniently into three parts:

1. The effect on the "effective" emission height.
2. The effect on transport of the pollutants.
3. The effect on the dispersion of the pollutants.

The effective height of pollutant sources

After an effluent leaves its source (such as a stack or chimney), it usually keeps rising. The more it rises, of course, the less pollution is produced at ground level. The amount of rise is controlled both by meteorological and nonmeteorological factors. The most important nonmeteorological factors are the area of the source and the efflux velocity. The effective height of the source also depends critically on the difference between the temperature of the effluent and that of the surrounding air, a quantity which, in part, depends on meteorological conditions. The most important purely meteorological factor is the wind speed, in the sense that a strong wind reduces the height to which the effluent will rise. Another

major meteorological factor (which is also extremely important in controlling dispersion) is lapse rate, defined as the decrease of temperature with height.

In the atmosphere below 10 km or so, the temperature usually decreases upward. However, there are layers of limited vertical extent in which the temperatures increase with height. Such layers are called "inversion layers" or, simply, "inversions." For example, inversions are common near the ground at night over rural terrain. Elevated inversions also occur frequently in some regions, occasionally in others.

The vertical change of temperature determines the hydrostatic stability (often just called "stability"). Inversion layers (warm air on top of cold) are most stable. They resist any kind of mixing or vertical penetration of effluent. If the lapse rate is positive and large enough (cold air over warm), there is a tendency for convection currents to be set up as the warm air below attempts to rise. In this state, the air is considered unstable.

To return to the problem of effluent rise, inversion layers limit the height and cause the effluent to spread out horizontally; in unstable air, the effluent theoretically keeps on rising indefinitely—in practice, until a stable layer is reached.

There exist at least 20 formulae which relate the rise to the meteorological and nonmeteorological variables. Most are based on fitting equations to smoke rise measurements. Because many such formulae are based only on limited ranges of the variables, they are not generally valid. Also, most of the formulae contain dimensional constants suggesting that not all relevant variables have been included properly.

For a concise summary of the most commonly used equations, the reader is referred to a paper by Briggs (1968). In this summary, Briggs also describes a series of smoke rise formulae based on dimensional analysis. These have the advantage of a more physical foundation than the purely empirical formulae and appear to fit a wide range of observed smoke plumes. For example, in neutrally stable air, the theory predicts that the rise should be proportional to horizontal distance to the $\frac{2}{3}$ power, which is in good agreement with observations. Although the dimensional-analysis formulae are not yet in general use, they appear to be the most promising in the long run.

Given the height of effluent rise above a stack, an "effective" source is assumed for calculation of tranport and dispersion. This effective source is assumed to be slightly upwind of a point straight above the stack, by an amount of the excess rise calculated. If the efflux velocity is small, the excess rise may actually be negative at certain wind velocities (downwash).

Transport of pollutants

Pollutants travel with the wind. Hourly wind observations at the ground are available at many places, particularly airports. Unfortunately, such weather stations are normally several hundred kilometers apart, and good wind data are lacking in between. Further, wind information above 10 meters height is even less plentiful, and pollutants travel with winds at higher levels.

Because only the large-scale features of the wind patterns are known, air pollution meteorologists have spent considerable effort in studying the wind patterns *between* weather stations. The branch of meteorology dealing with this scale—the scale of several to 100 km—is known as mesometeorology. The wind patterns on this scale can be quite complex, and are strongly influenced by surface characteristics. Thus, for instance, hills, mountains, lakes, large rivers, and cities cause characteristic wind patterns, both in the vertical and horizontal. Many vary in time, for example, from day to night. One of the important problems for the air pollution meteorologist is to infer the local wind pattern on the mesoscale from ordinary airport observations.

Moreover, pollutants are strongly affected by vertical motions which are sometimes produced by cities. No routine observations of vertical motions exist.

In many areas, local wind studies have been made. A particularly useful tool is the tetroon, a tetrahedal balloon which drifts horizontally and is followed by radar. Further, there exist theory and observations on sea breezes, mountain-valley effects, and so forth. In some important cities, such as New York and Chicago, the local wind features are well-known. In general, however, the wind patterns on the mesoscale are understood qualitatively, but not completely quantitatively. Much needs to be done in this area.

Atmospheric dispersion

Dispersion of a contaminant in the atmosphere essentially depends on two factors: on the mean wind speed, and on the characteristics of atmospheric "turbulence." To see the effect of wind speed, consider a stack which emits one puff per second. If the wind speed is 10 m/sec, the puffs will be 10 m apart; if it is 5 m/sec, the distance is 5 m. Hence, the greater the wind speed, the smaller the concentration.

Atmospheric "turbulence" consists of horizontal and vertical eddies which are able to mix the contaminated air with clean air surrounding it; hence, turbulence decreases the concentration of contaminants in the plume, and increases the concentration outside. The stronger the turbulence, the more the pollutants are dispersed.

There are two mechanisms by which "eddies" are formed in the atmosphere: heating from below and wind shear. Heating produces convection. Convection occurs whenever the temperature decreases rapidly with height—that is, whenever the lapse rate exceeds 1°C/100 m. It often penetrates into regions where the lapse rate is less. In general, convection occurs from the ground up to several hundred meters elevation on clear days and in cumulus-type clouds.

The other type of turbulence, mechanical turbulence, occurs when the wind changes with height. Because there is not wind at ground level, and there usually is some wind above the ground, mechanical turbulence just above the ground is common. This type of turbulence increases with increasing wind speed (at a given height) and is greater over rough terrain than over smooth terrain. The terrain roughness is usually characterized by a "roughness length," z_0, which varies from about 0.1 cm over smooth sand to a few meters over cities. This quantity does not measure the acutal height of the roughness elements; rather it is proportional to the size of the eddies that can exist among the roughness elements. Thus, if the roughness elements are close together, z_0 is relatively small.

The relative importance of heat convection and mechanical turbulence is often characterized by the Richardson number, Ri. Acutally, $-Ri$ is a measure of the relative rate of conversion of convective to mechanical energy. For example, negative Richardson numbers of large magnitude indicate that convection predominates; in this situation, the winds are weak, and there is strong vertical motion. Smoke leaving a source spreads rapidly, both vertically and laterally (Fig. 3). As the mechanical turbulence increases, the Richardson number approaches zero, and the angular dispersion decreases. Finally, as the Richardson number becomes positive, the stratification becomes stable and damps the mechanical turbulence. For Richardson numbers above 0.25 (strong inversions, weak winds), vertical mixing effectively disappears, and only horizontal eddies remain.

Because the Richardson number plays such an important role in the theory of atmospheric turbulence and dispersion, Table 1 gives a qualitative summary of the implication of Richardson numbers of various magnitudes.

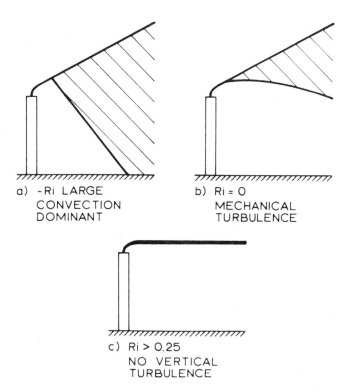

a) -Ri LARGE
CONVECTION
DOMINANT

b) Ri = 0
MECHANICAL
TURBULENCE

c) Ri > 0.25
NO VERTICAL
TURBULENCE

Figure 3. Spread of pollution from an elevated source for different Richardson numbers. (Schematic)

It has been possible to describe the effect of roughness length, wind speed, and Richardson numbers on many of the statistical characteristics of the eddies quantitatively. In particular, the standard deviation of the vertical wind direction is given by an equation of the form:

$$\sigma_\theta = \frac{f(Ri)}{\ln z/z_0 - \psi(Ri)} \qquad (1)$$

Here z is height and $f(Ri)$ and $\psi(Ri)$ are known functions of the Richardson number which increase as the Richardson number decreases. The vertical wind direction plays an important role in air pollution, because its functions determine the initial angular spread of a plume in the vertical. If it is large, the pollution spreads rapidly in the vertical. It turns out that under such conditions, the contaminant spreads rapidly sideways, so that the central concentration decreases rapidly downstream. If σ_θ is small, there is negligible vertical spreading.

Equation (1) states that the standard deviation of vertical wind direction does not explicitly depend on the wind speed but, at a given height, depends only on terrain roughness and Richardson number. Over rough terrain, vertical spreading is faster than over smooth terrain. The variation with Richardson number given in equation (1) gives the variation of spreading with the type of turbulence as indicated in Table 1: greatest vertical spreading with negative Ri with large numerical values,

less spreading in mechanical turbulence $(Ri=0)$, and negligible spreading in stable temperature stratification with little wind change in the vertical.

An equation similar to equation (1) governs the standard deviation of horizontal wind direction. Generally, this is somewhat larger than σ_θ.

In summary, then, dispersion of a plume from a continuous elevated source in all directions increases with increasing roughness, and with increasing convection relative to mechanical turbulence. It would then be particularly strong on a clear day, with a large lapse rate and a weak wind, particularly weak in an inversion, and intermediate in mechanical turbulence (strong wind).

Estimating concentration of contaminants

Given a source of a contaminant and meteorological conditions, what is the concentration some distance away? Originally, this problem was attacked generally by attempting to solve the diffusion equation:

$$\frac{d\chi}{dt} = \frac{\partial}{\partial x} K_x \frac{\partial \chi}{\partial x} + \frac{\partial}{\partial y} K_y \frac{\partial \chi}{\partial y} + \frac{\partial}{\partial z} K_z \frac{\partial \chi}{\partial z} \qquad (2)$$

Here, χ is the concentration per unit volume, x, y and z are Cartesian coordinates, and the K's are diffusion coefficients, not necessarily equal to each other.

If molecular motions produced the dispersion, the K's would be essentially constant. In the atmosphere, where the mixing is produced by eddies (molecular mixing is small enough to be neglected), the K's vary in many ways. The diffusion coefficients essentially measure the product of eddy size and eddy velocity. Eddy size increases with height; so does K. Eddy velocity varies with lapse rate, roughness length, and wind speed; so does K. Worst of all, the eddies relevant to dispersion probably vary with plume width, and depth, and therefore with distance from the source. Due to these complications, solutions of equation (2) have not been very successful with atmospheric problems except in some special cases such as continuous line sources at the ground at right angles to the wind.

The more successful methods have been largely empirical: one assumes that the character of the geometrical distribution of the effluent is known, and postulates that effluent is conserved during the diffusion process (this can be modified if there is decay or fall-out).

The usual assumption is that the distribution of effluent from a continuous source has a normal (Gaussian) distribution relative to the center line both in the vertical direction, z (measured from the ground) and the direction perpendicular to the wind, y. The rationalization for this

Table 1. Turbulence Characteristics with Various Richardson Numbers.

$0.25 < Ri$	No vertical mixing
$0 < Ri < 0.25$	Mechanical turbulence, weakened by stratification
$Ri = 0$	Mechanical turbulence only
$-0.03 < Ri < 0$	Mechanical turbulence and convection, but mixing mostly due to the former
$Ri < -0.04$	Convective mixing dominates mechanical mixing

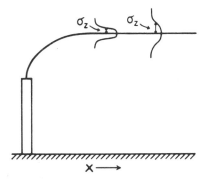

Figure 4. Vertical distribution of effluent as a function of distance from the source. (Schematic)

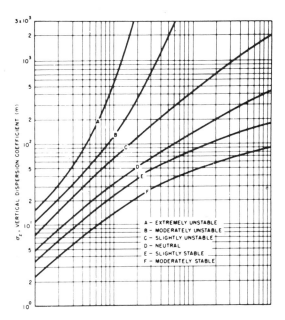

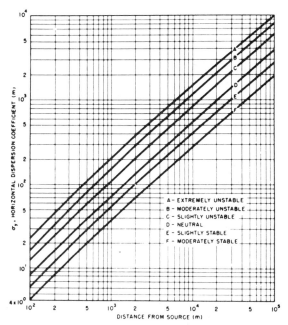

Figure 5. Nomograms for estimating horizontal and vertical spread as function of downwind distance and Pasquill's stability category.

assumption is that the distributions of the velocity components are also nearly normal. Subject to the condition of continuity, the concentration is given by (including reflection at the ground)

$$\chi = \frac{Q}{2\pi V \sigma_y \sigma_z} \exp - \frac{y^2}{2\sigma_y^2} \left[\exp - \frac{(z-H)^2}{2\sigma_z^2} + \exp - \frac{(z+H)^2}{2\sigma_z^2} \right] \quad (3)$$

Here, H is the "effective" height of the source, given by stack height plus additional rise, σ is the standard deviation of the distribution of concentration in the y and z direction, respectively, and V is the wind speed, assumed constant. Q is the amount of contaminant emitted per unit time.

The various techniques currently in use differ in the way σ_y and σ_z are determined. Clearly, these quantities change with downwind distance x (Fig. 4) as well as with roughness and Richardson number.

Quantitative estimation of the Richardson number requires quite sophisticated instrumentation; approximately, the Richardson number can be estimated by the wind speed, the time of the day and year, and the cloudiness. Thus, for example, on a clear night with little wind, the Richardson number would be large and positive, and σ's in equation (3) are small; on the other hand, with strong winds, the Richardson numbers are near zero, and the dispersion rate as indicated by the σ would be intermediate.

For many years, standard deviations were obtained by Sutton's technique, which is based on a very arbitrary selection for the mathematical form of Lagrangian correlation functions. More popular at present is the Pasquill-Gifford method in which σ_y and σ_z as function of x are determined by empirical graphs (Fig. 5). Note that the dependence of the standard deviations on x varies with the "stability category" (from A to F). These categories are essentially Richardson number categories, judged more or less subjectively. Thus, A (large dispersion) means little wind and strong convection; D is used in strong winds, hence strong mechanical turbulence and less dispersion; F applies at night in weak winds.

The main drawback of the Pasquill-Gifford method is that it does not allow for the effect of terrain roughness; the empirical curves were actually based on experiments over smooth terrain, and therefore underestimate the dispersion over cities and other rough regions. Some users of the method suggest allowing for this by using a different system of categories over rough terrain than originally recommended.

This difficulty can be avoided if measurements of horizontal and vertical wind direction variations are available; in general, the standard deviations of wind direction are proportional to the angular dispersion of contaminants, as already mentioned, no matter what the meteorological and surface conditions. For example, Smith (1968) suggests equations of the forms in stable air:

$$\sigma_y = 0.15 \, \sigma_\alpha x^{0.71}$$
$$\sigma_z = 0.15 \, \sigma_\theta x^{0.71}$$

and in unstable air: (4)

$$\sigma_y = 0.045 \, \sigma_\alpha x^{0.86}$$
$$\sigma_z = 0.045 \, \sigma_\theta x^{0.86}$$

Here, σ_α and σ_θ are the standard deviations of vertical and horizontal wind directions in degrees, and lengths are in meters.

Although these equations are not in general use, this approach seems quite promising, except for some controversies about the exact numbers to be used in equations (4).

As the plume expands vertically, the vertical distribution cannot remain normal indefinitely. At the bottom, the plume is limited by the ground. At the top, the plume will be limited by an elevated inversion layer. Eventually, the vertical distribution becomes uniform. In that case, the concentration is given by the equation:

$$\chi = \frac{Q}{\sqrt{2\pi}\ VD\sigma_y} \exp - \frac{y^2}{2\sigma_y{}^2} \qquad (5)$$

where D is the height of the inversion layer, which is also the thickness of the "mixed layer." Note that the concentration is inversely proportional to VD, the "ventilation factor," which is the product of D, and V, the average wind in the mixed layer.

The lateral spread is often limited by topography. In a valley of width W, the factor $(\exp - y^2/2\ \sigma_y{}^2)\ / (\sqrt{2}\ \pi\ \sigma_y)$ in equations (3) and (5) is replaced by $1/W$, after the contaminant concentration fills the valley uniformly in the y-direction (the direction perpendicular to the wind, and also usually perpendicular to the valley). The effect of this change is that relatively large concentrations are maintained at large distances from the sources.

City models

The equations above given the pollutant concentrations downwind from a single source. In order to obtain the total picture of air pollution from a city, the concentrations resulting from all sources must be added together, for all different wind directions, different meteorological conditions, and for each contaminant. Such a procedure is expensive, even if carried out with an electronic computer, and even if, as is usually done, all small sources in square-mile areas are combined. Therefore, complete city models of air pollutant concentrations have only been constructed for very few locations. It is necessary, however, to have city models in order to understand the distribution of contaminants; only then it is possible to determine the most economical strategy to reduce the pollution.

Because the construction of a complete city model is so expensive, city models are often simplified. For example, if the city is represented by a series of parallel line sources, the computations are greatly reduced. Many other simplifications have been introduced; for a summary of most city models now in existence, see Stern (1968).

Diurnal variation of air pollution

Equation (5), which shows that concentrations at considerable distances from individual sources are inversely proportional to the ventilation factor (VD), can be used to explain some of the variations in air pollution caused by meteorological factors. First, we shall consider the diurnal variation of air pollution. Of course, the actual variation of pollution may be different if the source strength varies systematically with time of day. The diurnal variation is different in cities and in the country. Consider typical vertical temperature distributions as seen in

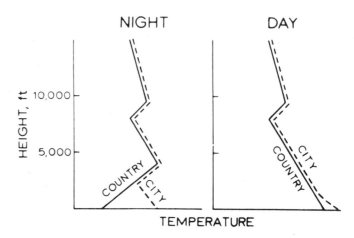

Figure 6. Schematic vertical temperature distributions.

Figure 6. During the day, both over cities and country, the ground temperature is high, giving a deep mixed layer. After sunset, the air temperature near the surface in the country falls, producing an inversion reaching down to the ground. After air moves from the country out over the relatively warmer and rougher city, a thin mixed layer is formed near the ground. The thickness of this mixed layer varies with the size of the city, and depends on how long the air has moved over the city. In New York, for example, the mixed layer is typically 300 m thick; in Johnstown, Pa., an industrial valley city with just under 100,000 population, it is only a little over 100 m.

Figure 7 indicates how the temperature changes shown in Figure 6 influence the diurnal variation of pollution due to an elevated source in the country; at night, vertical mixing is negligible and the air near the ground is clean. Some time shortly after sunrise, the mixed layer extends to just above the source, and the elevated polluted layer is mixed with the ground air, leading to strong pollution (also referred to as "fumigation"), which may extend many kilometers away from the source. Later in the morning and early afternoon, the heating continues and thickens the mixed layer. Also, the wind speed typically increases, and the pollution decreases.

In the city, many sources usually exist in the thin night-time mixed layer. Since this layer is so thin, and the wind usually weak, large pollution occurs at night. Right after sunrise, the pollution at first increases somewhat, as the effluent from large, elevated sources is brought to the ground. As the mixed layer grows, the concentrations diminish, and, in the early afternoon, they are often less than the night-time concentrations (see Fig. 8).

Thus, the main difference between air pollution climates in the city and country is that country air near industrial sources is usually clean at night, whereas the city air is dirtier at night than in the middle of the day. These differences are most pronounced during clear nights and days, and can be obliterated by diurnal variations of source strengths. Figure 8 shows the characteristic behavior only because the sources of pollution at Johnstown, Pa., are fairly constant throughout the day.

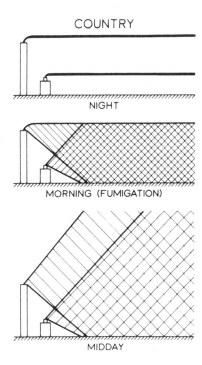

COUNTRY

NIGHT

MORNING (FUMIGATION)

MIDDAY

Figure 7. Night-to-day variation of pollution in the country. (Schematic)

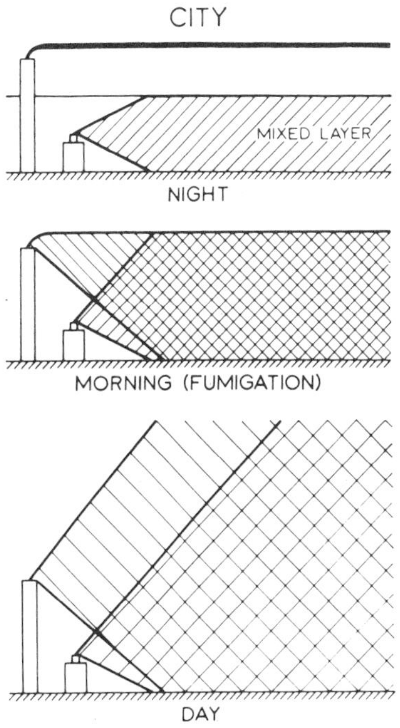

CITY

NIGHT

MORNING (FUMIGATION)

DAY

Figure 8. Night-to-day variation of pollution in the city. (Schematic)

Day-to-day variations in air pollution

Equation (5) shows that, other things being equal, the concentration of contaminants is inversely proportional to the wind speed. Figure 9 shows this effect on 24-hr total particulate concentration at Johnstown for cases where the source strengths were roughly the same, during the fall of 1964.

Conditions of particularly bad air pollution over wide areas and for extended periods are accompanied not only by light winds and calms, but also by unusually small mixing depths (D) so that the ventilation factor is unusually small. Such conditions occur within large high-pressure areas (anticyclones). In such areas, air is sinking. Sinking air is warmed by compression. Thus, in an anticyclone (high-pressure area), an elevated warm layer forms, below which there is room only for a relatively thin mixed layer (Fig. 10). The inversion on top of the mixed layer prevents upward spreading of the pollution, and when mountains or hills prevent sideways spreading, the worst possible conditions prevail. A particularly bad

situation arose in the industrial valley town of Donora, Pa., in which many people were killed by air pollution in 1949.

Cities in California, like Los Angeles, are under the influence of a large-scale anticyclone throughout the summer, and an elevated inversion at a few hundred meters height occurs there every day; that is why Los Angeles had air pollution problems as soon as pollutants were put into the atmosphere to any large extent. In the United States outside the West Coast, stagnant anticyclones occur only a few times per year, usually in the fall.

In order to warn of generally bad air pollution conditions, the Weather Bureau computes daily ventilation factors (VD), the depth of the mixed layer multiplied by the average wind speed within it, at all stations which make upper-air observations. These quantities are transmitted along with the usual weather data over the weather teletype network. If the ventilation factors remain low for 24 hours (and when some other conditions are satisfied), an air pollution alert is declared, at which

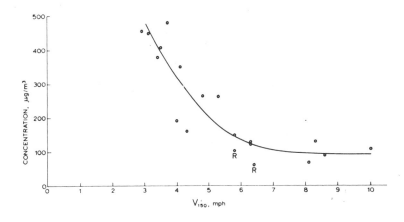

Figure 9. Variation of 24-hour average concentration of particulates at Johnstown, Pa., as function of wind speed at 150 ft. R indicates precipitation.

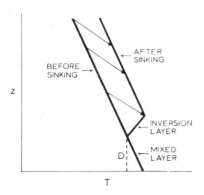

Figure 10. Development of elevated inversion by large-scale sinking, particularly in an anticyclone. (Schematic)

time the air pollution potential is expected to be high. On the basis of ventilation factors and air pollution alerts, some industries voluntarily cut their sources of contaminants.

Not much has been said about the influence of wind direction on air pollution. When pollution is mainly due to many, relatively small sources, as it is in New York, the pollution is surprisingly insensitive to changes in wind direction. Even in Johnstown, Pa., wind direction is unimportant except for the case of easterly winds, when a single, huge steel plant adds significantly to the contaminant concentration.

In contrast, wind direction plays a major role when most of the pollution in a given area is due to a single or a few major plants. Also, there are special situations in which wind direction is particularly important; for example, in Chicago, which has no pollution sources east of the city, east winds bring clean air.

The main difference between the effects of lapse rate, mixing depth, and wind speed on the one hand, and wind direction on the other, is that the wind direction has different effects at various sites, depending on the location of the sources; the other factors have similar effects generally.

Summary

This paper has given many of the meteorological factors on air pollution, both qualitative and quantitative. Many features have been omitted, such as fallout and deposition, instantaneous sources, change of wind with height, and special problems with radioactive pollution. Nevertheless, it is perhaps true to say that meteorological effects are rather well understood. It also seems true that this knowledge is not being used to the fullest extent at this time in attempts at air pollution abatement and in locating new industries.

Acknowledgments

The author is grateful to the following scientists for critically reading this manuscript and for providing constructive suggestions: R. A. McCormick, W. Moroz, D. Pack, I. Van der Hoven, and E. Peterson.

References

Briggs, Gary A., 1968: Meteorology and Atomic Energy, Atomic Energy Commission, Federal Clearinghouse No. TID-24190, p. 189.

Smith, Maynard, 1968: ASME Guide for Prediction of the Dispersion of Airborne Effluents, p. 54.

Stern, Arthur, 1968: Air Pollution and Its Effects, McGraw Hill Book Company, Vol. 1, p. 220.

John T. Middleton

Views

Planning against Air Pollution

*The Acting Commissioner of the Air Pollution Control Office
discusses the major sources of air pollution and methods
used for its measurement and control*

Our environment is changing—we are changing it—and serious questions are being raised as to whether man can adapt to some of these changes. Our air for example is heavily burdened with dirt and chemicals. It constricts our throats and makes us sick; it kills some of us ahead of our time. It besmirches our great cities and destroys some of the material goods and pleasures for which we began enduring the pollution in the first place.

We could solve this problem instantly by stopping our cars, closing our industries, and giving up many of the things that we believe are worthwhile. In time the air would purge itself of most of its impurities and we could live primitively and pastorally in a dull, uncomfortable Walden. But most of us would prefer something else. And so, as a matter of national policy, government at several levels is working within its financial and professional capabilities to limit pollution.

The motor vehicle

The automobile is at present the source of almost half of all air pollution in this

An international authority in the broad field of air pollution, Dr. John T. Middleton was first recognized for his early detection of photochemical air pollution as an adverse economic factor to California agriculture in the mid-1940s. After a career on the plant pathology faculty of the University of California (Riverside and Los Angeles), during which he also was for ten years director of the University of California Statewide Air Pollution Research Center, he was appointed on January 1, 1967, to be director of the newly created National Center for Air Pollution Control, under the Department of Health, Education and Welfare, and has headed the national air pollution control program ever since. He is now Acting Commissioner of the Air Pollution Control Office in the Environmental Protection Agency. Dr. Middleton is a member of numerous professional associations and has served on many government boards. Address: Environmental Protection Agency, Air Pollution Control Office, 5600 Fishers Lane, Rockville, MD 20852.

country. It accounts for more than half the hydrocarbons, nearly half the nitrogen oxides, and two-thirds of the carbon monoxide that are released into the air of the United States each year. It is also the chief source of lead in the atmosphere. The hydrocarbons are a vast family of chemicals, most of which are directly harmful only in high concentrations as gases. However, some hydrocarbons are toxic in themselves, and others participate in a series of reactions with the oxides of nitrogen that takes place in the presence of sunlight. This leads to the formation of new, highly troublesome and toxic products called "oxidants," which cause eye irritation, allergenic responses, reduced work performance, depressed mitochondrial activity, altered functioning of several biological systems, and destructive effects on vegetation, polymers, cellulose, and dyes.

Even quite low levels of oxidant pollution, on the order of 200 to 590 $\mu g/m^3$ (0.05 to 0.30 ppm), have been shown to cause eye irritations and to impair the performance of young athletes. Eight-hour exposure to carbon monoxide concentrations that are common in cities has been shown to produce temporary impairment of mental performance.

We have little information on the effects on man's health of exposure to the levels of nitrogen oxides commonly found in urban air. However, a study made in Chattanooga, Tennessee, has tentatively linked low levels of these oxides to children's susceptibility to respiratory illness.

We do not know the significance of urban smog in geophysical changes caused by pollution on a global scale. But photochemical smog is known to

affect city climates, and it appears to be responsible for widespread chronic and acute damage to vegetation hundreds of miles from urban centers, so the possibility of secondary long-term effects must be considered.

Finally, there is lead. Emissions of lead into the nation's air are now about 200,000 tons per year. Ten years ago the total was about 130,000 tons. Symptoms of lead poisoning in humans have been reported at blood levels as low as 60 micrograms of lead per 100 grams of blood. It has been shown that persons who are frequently exposed to lead, such as traffic policemen, have more of it in their blood than other persons; and the normal blood levels of persons who are not unusually exposed now range from 10 to 30 micrograms of lead per 100 grams of blood. This is not much of a safety factor in view of the increase in emissions that expose more and more of us to the hazard. Recognizing this fact and the need to provide a new motor vehicle emission-reduction technology, the petroleum industry has shown that the lead can be removed.

The Environmental Protection Agency (EPA) has been requiring progressively more stringent pollution controls on automobiles. First it required the elimination of hydrocarbon emissions from the crankcase and partial control of exhaust hydrocarbons in new models for 1968. Then it tightened the exhaust and carbon monoxide standards for 1970 cars. The 1971 models also had to meet the first standards limiting the evaporative loss of hydrocarbons from the engine and fuel system. The cars of 1972 will have to meet new standards for evaporative hydrocarbons and will be tested under far more exacting conditions than any that have been used heretofore, to achieve a 69 percent reduction in carbon monoxide and an 80 percent reduction in hydrocarbons. The first limit on emissions of oxides of nitrogen will go into effect in 1973.

The 1970 Clean Air Amendments provide for the testing of production-line models of automobiles, as well as the prototypes that were tested under the 1967 Clean Air Act. The Amendments also provide for the reduction of 1975-model carbon monoxide and hydrocarbon emissions to a level that is 90 percent lower than was allowed on 1970 prototypes, and for the reduction of 1976-model emissions of nitrogen oxides to a level that is 90 percent below those that were reported in 1971-model cars. Reductions of this order, however, may not by themselves be adequate to achieve healthful air quality levels in the most congested urban areas; other steps and further emission reductions may be necessary.

To stimulate inquiry and uncover possible new approaches to the problem, the EPA began a Federal Clean Car Incentive program. It is hoped that this program will attract the talent and energy of a large number of inventive people and forward-looking companies, and lead to the production by the late 1970s of a pollution-free passenger car. Such a car should equal the convenience and performance of the present-day automobile and allow the matching of fuels to vehicles for better air pollution control. And all of this is an add-on—not a substitute for mass transit.

Industrial pollution

Let us turn to the effort to reduce emissions from stationary sources—mainly industrial plants. Since the Clean Air Act of 1967, we have been drawing geographical boundaries around air quality control regions for those places where air pollution is a particularly serious problem. We have so far designated 101 such regions and will increase that number as the states request that additional areas be designated as set out in the Clean Air Amendments of 1970.

We have published criteria reports to tell what science has thus far been able to reveal of the insidious as well as the obvious effects of air pollution on man and his environment. So far, these reports cover the five most important classes of pollutant—sulfur dioxide, particulate matter, carbon monoxide, hydrocarbons, and photochemical oxidant. With each of these reports, we have published also a summary of methods for controlling these pollutants.

It has been up to the States with federal help to devise, adopt, and implement regional air quality standards that would safeguard the public health. But under the time schedule set forth in the Clean Air Act, only 26 States submitted standards, and only 17 regional implementation plans designed to achieve those standards have been submitted.

Now the Clean Air Act has been amended to provide for faster action. By February 1st, we will issue proposed national air quality standards on the pollutants that I have already mentioned, plus another—oxides of nitrogen—for which the criteria and technology reports also will be issued at the same time.

These standards will be based on the criteria documents, which represent the best evidence presently attainable on the effects of air pollution. To protect the public health, there will be nationwide primary air quality standards which will define how clean the air must be in order to be healthful to breathe. To safeguard the public welfare, there will be nationwide secondary air quality standards, which will say how clean the air must be in order to protect us against the known or anticipated effects of air pollution on property, materials, climate, economic values, and personal comfort.

It is significant, I think, that these standards will be set entirely on the basis of the need to protect public health and welfare, and the need to attain them will either force the use of available technology or create new pressure for the discovery and demonstration of new methods to control air pollution; we will set the goals and it will be up to scientists, technicians, and planners to find ways to meet them.

Under the 1970 amendments, the states will continue to have primary responsibility for devising regulatory and enforcement procedures to achieve the necessary improvements in air quality. This is work that must begin at once; it must reflect the kind of social and political decisions that are inherent in reforms of this magnitude. Public participation in this process is encouraged by law. As a result, what the *Wall Street Journal* has called a "breathers' lobby" has developed to press for cleaner air across the country.

The new law also provides for the establishment in 1971 of federal performance standards for new stationary sources of air pollution, reflecting the use of the best system of emission reduction that has been adequately demonstrated, and for the establishment of federal emission limitations for hazardous pollutants that may cause or contribute to an increase in mortality or an increase in serious irreversible or incapacitating illness. New sources, let me make clear, are those that are built after the date of any proposed standards, or modified in such a way as to emit any pollutant not previously emitted or cause an increase in present pollutants.

The states may also be empowered to administer these provisions if their clean-air officials have authority to review plans and specifications of new industrial facilities, to prohibit the construction of those that fail to meet the national standards, and in the case of hazardous pollutants, to provide adequate and speedy enforcement procedures to safeguard public health.

So far, progress in the control of pollution from stationary sources is furthest advanced in the control of sulfur oxides and particulate matter. The sulfur oxide problem is particularly urgent because the pollutant is so hazardous, so common, and so closely linked to vital industries. More than half of all sulfur oxide emissions in this country come from electric generating plants. Nationally, approximately 37,000,000 tons of sulfur oxides will be discharged into our atmosphere this year. Unless adequate control measures are taken, our most optimistic forecasts indicate that these emissions will increase over 60 percent by 1980 and almost fourfold by the year 2000.

National energy policy

The issue of control hinges, in the short term, on the development of a wise energy policy for the nation, a policy which will reflect our concern with social and environmental objectives as well as promote economic progress. The creation of such a policy should be a matter of pressing concern

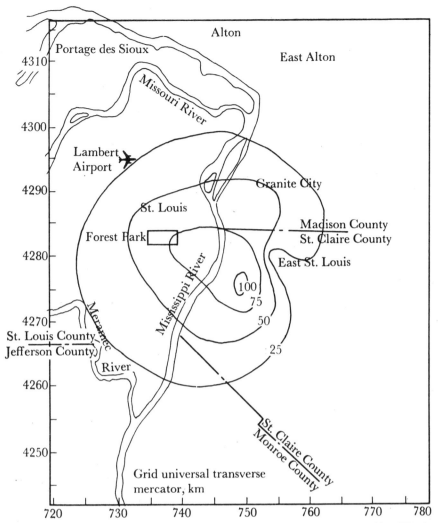

Figure 1. Observed sulfur oxide concentrations (annual arithmetic mean, $\mu g/m^3$) in 1967 in East St. Louis–Granite City, Illinois, and St. Louis, Missouri.

to the scientist both as scientist and citizen. To reduce the air pollution from oxides of sulfur, particulates, and oxides of nitrogen by using other energy-producing techniques will not solve the overall environmental problem if new types of air pollution problems are generated, and if no solution is found for thermal pollution. In the case of nuclear generating plants, it is assumed that environmental problems will be resolved so that we will be able to use as much atomic power as we now expect. The use of solar cells and fuel cells to generate electricity could, by the 1980s, be increasingly important factors in solving the nation's total energy problem with a minimum of environmental pollution.

Attention is already being given to attempting to define the possible implications of various fuel policy options. The Federal Power Commission, the Atomic Energy Commission, the Office of Science and Technology, and

the Office of Energy Preparedness will certainly influence any final decisions, as will the Department of Interior, the industries affected, and the public. Certainly the voices of the Council on Environmental Quality and the EPA will also be heard.

I cannot attempt to predict what will finally emerge when these various viewpoints have been weighed. But I can mention some of the considerations of the EPA.

It is becoming more evident all the time that the United States lacks adequate reserves of gas and oil to supply indefinitely our growing demand for energy. Coal is the only readily available domestic fossil fuel with reserves that are large enough to fill such needs for generations to come; but most of the coal in the country contains large quantities of sulfur, which are changed into sulfur oxides by the combustion process. We may have to take steps to

get our reserves of low-sulfur coal and oil developed preferentially. If we adopt a goal of obtaining within ten years adequate supplies of low-sulfur gas from coal, we can perhaps take a less restrictive view toward problems of depleting reserves of natural gas.

We may have to consider providing incentives for importing more liquefied natural gas and developing more liquid natural gas storage facilities. Environmental problems and increasing costs of fuel make importation of liquefied natural gas seem to be probable future steps.

The EPA's present approach within its own purview is to couple state regulation of emissions with the development of technology in order to provide alternate solutions to reconcile our energy demands with the demand for a clean environment. We have to look for new technology in order to assure that air quality standards will be attainable in a reasonable time.

While a national energy policy is being developed and applied, we should be developing, testing, and providing techniques to remove sulfur from fuels and the products of combustion. Inquisitive and talented men and women are needed to develop more efficient combustion techniques by the end of the decade and to determine whether or not our limited oil supplies are to be used for chemical production or for heat. What do we do with our abundant coal supplies—gasify them? Is it better resource management to produce sulfur as a reclamation product of combustion or to mine it?

Sulfur oxides

Unless scientists and others apply themselves to answering these questions, we shall continue to make ourselves sick with sulfur dioxide. A study in Rotterdam showed a positive association between total mortality and exposure for a few days to 24-hour average concentrations of 0.19 ppm of sulfur dioxide; there was an indication that the association held true even at lower levels. These levels are often exceeded in many cities in the United States. A major British study found an association between air pollution and deaths from bronchitis and lung cancer in an area where the yearly average level of sulfur dioxide was 0.04 ppm;

this level is quite common now in our cities.

Solid or liquid particulate matter in the air can carry sulfur dioxide deep into the lungs, causing injuries much more severe than are encountered in laboratory experiments using sulfur dioxide alone. The health hazards of air polluted with sulfur dioxide and particulates have been demonstrated in a number of air pollution disasters; for example in Donora, Pennsylvania, in New York City, and in London.

In the atmosphere sulfur dioxide reacts rather quickly with other materials to form sulfuric acid and salts of sulfuric acid. From the geophysical point of view, the EPA is studying the formation and persistence of sulfate aerosol; from the biological point of view, we are looking at the accumulation in soil and surface waters of sulfuric and sulfurous acids and their salts following rain and fallout. There is some evidence that over some seas and tropical land areas the condensation nuclei are largely ammonium sulfate; and the sulfate aerosol is estimated to persist for days in the troposphere and for hundreds of days in the stratosphere.

On a local scale, the sulfur family of pollutants corrodes metals, disintegrates paints, causes fibers to weaken and fade, discolors building materials and makes them deteriorate. Agricultural production drops as plant growth and yield are suppressed. And sulfur oxides contribute to the reduction of visibility that often accompanies air pollution.

Particulate pollution

By itself, undifferentiated particulate pollution—from particles under 500 microns in diameter—can injure surfaces within the respiratory system and affect climate, visibility, building materials, textile fibers, and vegetation. Specific substances that may become suspended in the air as particulate matter of course require individual assessment. One of the most common, and the only one I shall deal with here, is lead. Not only are human-body burdens of lead rising near points of peril; the total environment is accumulating lead. Particles of lead from air pollution—most of them originating in the automobile, which spews out 90 percent of its lead particles at sizes smaller than two-tenths of a micron—have been found in the snow

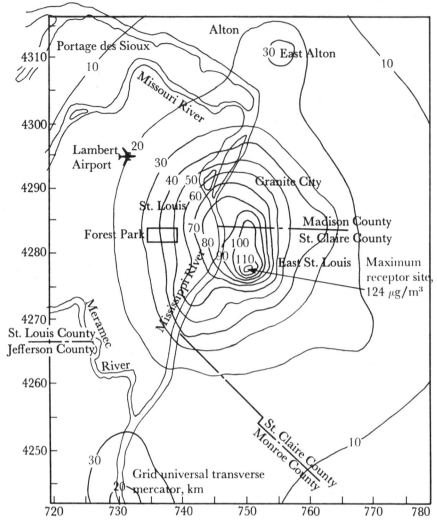

Figure 2.   Estimated sulfur oxide concentrations based on simulation modeling from 1967 emissions data.

of polar ice caps. In these layerings of one winter's snow upon the next it is possible to discern the annual increase in the earth's airborne lead pollution.

As we approach the danger point for human beings, are we approaching the danger points for other forms of life? I do not know, but I do believe that lead, once it is in the environment, is an ominous and persistent threat, narrowing our margins of life-safety.

Weather as a factor

Weather is a vital factor in the pollution problem that is of particular interest to the scientist. On the one hand, we may be raising the earth's temperature with carbon dioxide that blocks the long-wave radiation of heat away from our planet; on the other hand, we may be cooling our earth by blocking solar energy with a great shield of particulate pollution. And not only do pollutants affect the weather in various

other ways—as by nucleating clouds and producing acidic rainfall—but there are reciprocal effects. Weather conditions, as I have said, affect the transformation of automotive emissions into photochemical smog; and they can clamp the smog tightly upon the communities where it is produced.

Ordinarily, thermal and mechanical turbulence mix and dilute pollutants in the atmosphere; and the wind, if it is strong enough, can scatter them. Ultimately, natural processes can remove man-made pollutants from the atmosphere. Some, in time, escape into space. Rain and snow eventually wash out the rest—whether in solid, liquid, or gaseous form—and deposit them in the earth's soil and water. But when the surface air is cooler than a layer of air above it, and mixing cannot occur, the atmosphere is said to be stable—the condition is known as an inversion—and pollutants accumulate in the limited air space. At one time

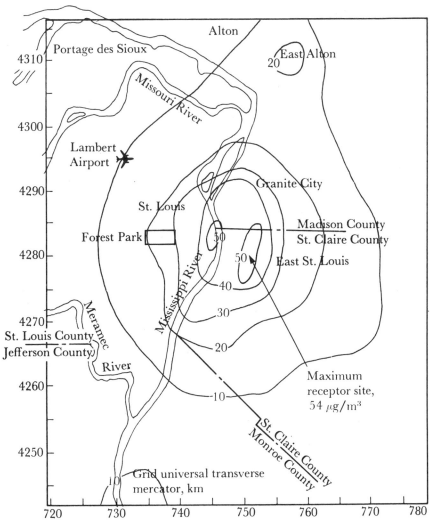

Figure 3. One emissions strategy, tested through simulation modeling, reduces estimated SOx pollution by 30 percent.

modeling involving sophisticated use of computers and plotting techniques. Only through the use of a simulation model can air pollution control be advanced from the haphazard to the systematic.

Let us say that the state or local control agency has measured a number of different concentrations of sulfur dioxide in St. Louis (Fig. 1). (Such measurements have already been made in a number of air quality control regions across the country.) The control officer needs to know how stringently to limit this or that source or group of sources in order to achieve healthful air at the point where pollution is worst.

Air pollution concentrations in our simulation model are a function of rates of emission, rates of pollutant decay, the height at which the pollutant is released into the air, its temperature at emergence, the topography of the community, and weather conditions. The related variables provide a basis for drawing on a map of the community isopleths of concentration from a source or group of sources. The simulation model (Fig. 2) will fill in the gaps left by local control agency measurements.

Most local sampling networks are not sufficiently extensive to identify all heavily contaminated areas or to indicate in any detail the differences in levels of pollution. The estimates provided by the model supplement the observed data and furnish much of this detail. This added information is subject to confirmation by comparison with the data gathered directly from the air sampling network. We see that there is a high degree of correlation between Figures 1 and 2. Also, the simulation indicates the need for additional sampling to the north of present sites; in this way it can be used as a basis for appropriate expansion and rearrangement of the air sampling network.

Further, the simulation provides the control officer with a ready tool for theoretical testing of various limits that might be selected for different sources or categories of sources. These limits—called emission standards—would be put into effect through legally enforceable control regulations.

By using various emission standards in the model, reductions in pollution levels ranging from around 30 to 80

Antipollution strategies

The design of strategies to reduce hazards to health from air pollution hinges on an ability to predict phenomena of this nature precisely enough and well enough in advance so that positive emergency actions may be taken and the availability assured of adequate technologies to control the sources of pollution.

Some of us remember the dismal, dark, black, coal smoke days of the 1930s. Some progressive urban areas have traded those old black days for new whiskey-brown days of photochemical smog.

Let us use St. Louis as a case study. Wind currents from East St. Louis on some days transport the Illinois con-

tribution to air pollution and add it to the Missouri problem. On other days, of course, Missouri gives it back. In the St. Louis air quality control region, the airborne burden of sulfur oxides will have to be reduced by 81 percent in the next few years in order to meet Missouri air quality standards. The present burden of particulate matter there will have to be reduced 61 percent. The improvements called for in Los Angeles under the California standards are 64 percent for sulfur oxide and 54 percent for particulates. For Cleveland they are 70 percent each. These are not measurements of the efficiency of control devices; these are the reductions needed in ambient pollution in order to meet the state standard of air quality.

Simulation modeling

The task of relating such goals to the design of emission controls to achieve the goals is a function of simulation percent can be achieved (Figs. 3, 4, 5). Some cities might not need to achieve as great reductions as other cities and might select different emission strategies. What is important is that intelligent choices be available.

Somewhat different approaches are used to set up simulation models for individual plants or factories—"point sources"—and area sources, such as a residential neighborhood having many minor sources of air pollution, or a congested highway; but the modeling procedure generally is similar. Also the model has limitations: it is more accurate over large areas than over areas the size of, say, a city block; and it is more accurate over the long term than the short. It is possible, however, to use the model to estimate short-term and neighborhood concentrations, if geometric standard deviations are used carefully.

The possibilities for the use of a tool such as the simulation model are not restricted to the control agency, nor are they restricted to the United States alone. The North Atlantic Treaty Organization—an agency set up primarily for the common defense—is attempting to use its technical resources to assist the Committee on Challenges to Modern Society to develop emission-reduction strategies for Frankfort, in Germany, and Ankara, in Turkey. NATO is using simulation modeling to achieve this goal, which is related to one of its lesser known but highly significant purposes: social and economic improvement in member countries.

Wherever the simulation model is used, it not only lends itself to the solution of immediate problems offering a variety of choices to correct conditions that exist today, but also provides insight into the problems that may exist in the future, when the community will have been further changed by growing urbanization and industrialization.

In the long term, strategies for emission reduction will have to take account of the changes projected in air quality due to changes in population trends, transportation, marketing, manufacturing, employment, energy requirements, solid-waste disposal needs, and the like—questions that have to do with the proper design of communities. Some communities

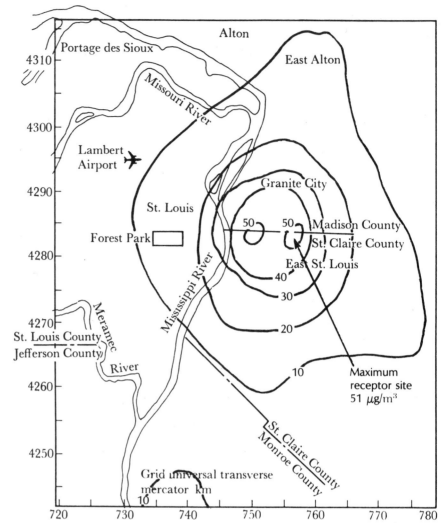

Figure 4. Simulation modeling reveals that another strategy would achieve a slightly greater reduction.

doubtless will discover that some neighborhoods—and some regions—simply cannot tolerate any more of certain kinds of industry, even after the best available technology is applied to that industry's pollution problems. Such situations may not be too far in the future for some urban areas.

It is also quite possible that simulation modeling may indicate the need for changes in highway planning or traffic patterns in order to reduce the hazard to the health of residents of some particular neighborhood. Given a continuation of present automobile marketing trends, such considerations will become of greater urgency to city planners and elected officials; in fact, there are some neighborhoods in this country where such problems are already causing concern.

These will not be easy decisions for the responsible authorities. One can imagine the consternation of shopkeepers if traffic past their shopping center had

to be rerouted during certain peak hours in order to avoid dangerously high levels of carbon monoxide or lead pollution in nearby suburbs. A high level of statesmanship, matching our concern for the public health, will be required of local officials.

Chicago plan for "emission rights"

One plan, under study in Chicago, would use the zoning ordinance as a device to control legally the effective emission density associated with each parcel of land in the region. This would limit the use of land to the extent necessary to achieve the regional air quality standards. The owner of a given parcel of land would own certain specific "emission rights" that are attached to the land until it is rezoned. Unused emission rights associated with one parcel of land might be sold to the owner of a neighboring parcel if he needed them.

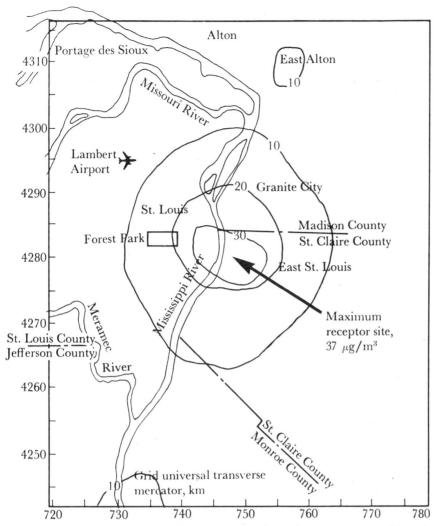

Figure 5. A third strategy would achieve an 80 percent reduction; officials may choose among the various strategies to achieve the result they desire.

The Chicago approach could be modified as required for emission controls based on global as well as regional (or zoning) perspectives. The global concern would reflect the general need to reduce pollution in order to limit the long-term buildup in the atmosphere. The regional perspective relates to the need to limit shorter term dosages as measured at ground-level receptor points. Global considerations, for example, might require all plants to reduce particulate emissions by 80 percent, regardless of stack height and location. The regional rule, however, might require more stringent levels of control for plants situated in specific neighborhoods.

Whatever is finally decided upon, the Chicago approach is at least interesting and reflects the broad-scale rethinking of pollution-control efforts and an awareness of the relationship between land use and pollution. President Nixon accented this relationship in his first Environmental Report to the Congress last August: "The uses to which our generation puts the land can either expand or severely limit the choices our children will have. ... Society as a whole has a legitimate interest in proper land use." The President's message was the first clear-cut national policy statement in our history on the prospect of land-use reform as a tool to achieve a more livable environment. The 1970 Clean Air Amendments insist on the use of that tool whenever it is needed to meet the national standards.

The broader role of the scientist

The scientist too has a new and vital role to play. In the past he was relied upon to enlarge the range of options that were available to the policymaker, but now those two functions are coming together. The needs of the time summon those who have special knowledge to use their influence when policy decisions are being made, so that the decisions will be more likely to be correct.

It is a difficult task that we are thrusting upon the scientist, but it will be even more difficult to evade. That disinterested impartiality so necessary for scientific inquiry may no longer, in these times, be cited to excuse the scholar from participation in the decisions of the nation. Such aloofness has nearly always been distrusted in the past, and in a social sense this disapproval is very healthy because it is based on deep instincts for the welfare of the nation as a whole.

Scientists must prepare themselves to meet this new responsibility and apply their special knowledge to the general welfare. The old and rigid disciplines of academic life are in need of cross-pollination. This is the time of the specially trained generalist, the man of many parts.

Our young people, many of them, see this. They are resisting the inflexible categorization of the outmoded university system. Help me become, they say, not a chemical engineer but a citizen engineer; not a chemist or physicist or biologist but a specially trained and highly aware citizen with responsibilities and tasks that cut across the old professional boundaries. The United States in the last half of the twentieth century desperately needs the participation and services of such universal men, for we are in a pretty pickle and our principles as well as our people are in danger. Only with the devoted effort, the courage, and the counsel of skilled and responsible scientists and technicians can America discover the options and select the best among them in order that our tomorrows may be as bright as our yesterdays seem to have been.

The Congress, in the National Environmental Policy Act of 1969, summoned each one of us to the task of preserving what we have left and repairing what we have lost. Our land and air and water, and the animal and plant life that sustains our human lives, are in our own hands. We as scientists have some of the knowledge of how to protect them and the responsibility for educating and training the new generations who will join the quest for a livable environment.

Paul F. Fennelly

The Origin and Influence of Airborne Particulates

Both man-made and natural processes contribute to the concentration of particulate matter in the atmosphere; the effects of this aerosol on human health, environment, and climate are just beginning to be understood

When we think about the air around us, we usually think of gases: oxygen, nitrogen, argon, carbon dioxide, water vapor. Yet the air also contains a variety of microscopic particles which, despite their exceedingly small size and low concentration, can exert a significant influence on atmospheric behavior and on human health and property.

Microscopic particles in the air, commonly called particulates, are not a new phenomenon. The haze characteristic of rural areas like the Smokey Mountains of North Carolina or the Blue Ridge Mountains of Virginia, for instance, is caused by airborne particulates. The particulates form after certain vapors emitted from the trees absorb sunlight and undergo a series of photochemical reactions in the atmosphere. The haze or coloration, caused by the manner in which the particles scatter sunlight, is part of the natural cycle, and no adverse effects are usually attributed to it.

A more recent phenomenon is the appearance over the last twenty-five years or so of intense hazes in our cities and industrial areas—the so-called urban smog (Fig. 1). The adverse effects of smog are common knowledge: dirt, grime, reduced visibility, and, in many cases, increases in morbidity and mortality rates. The accumulation of these adverse effects eventually triggered the passage of air pollution control legislation which set limits on the concentrations of various species in the air.

One of the targets of this legislation was to control the amount of particulate matter in the air. A standard of 75 micrograms per cubic meter was established as an acceptable upper limit for the annual average of total suspended particulates in the atmosphere in any given region (roughly 75 parts per billion by weight of dry air). For any given 24-hour period, a concentration of $260 \ \mu g/m^3$ was established as a maximum tolerable limit.

To achieve these standards, limits were set on the amount of solid material that could be discharged from various industrial smokestacks or automobile exhausts. Standards were also established for emissions of the gases SO_2, NO_2, O_3, CO, and hydrocarbons, which can cause adverse effects by themselves but are also of interest here because they contribute to the formation of particulates in the atmosphere.

In many urban areas, the air quality standards for suspended particulate matter are not being met, and there are those who think that the procedures for attaining these standards may be inadequate. On the other hand, some scientists are questioning whether the established standards are a realistic goal. There is some evidence that natural and uncontrollable sources of particulate matter in many cases contribute to the atmospheric loading in quantities which make the present standards unattainable. The situation is further complicated by the possibility of long-range transport of airborne particulates. A mounting body of data indicates that particulate matter or gases leading to the formation of particulate matter which are injected into the atmosphere in one location can be deposited at locations up to several hundred miles away. The implication is obvious: in regions where deposition or fallout occurs, the enforcement of strict pollution control procedures on a local basis will have little impact on air quality. The debate over standards for airborne particulate matter will probably intensify in the next few years.

Particulates are minute solid particles or liquid droplets ranging in size from 0.005 to about 500 microns. (A human hair, by comparison, is about 100 microns thick.) The size limits are rather arbitrary but are meant to indicate that the particles can be as small as a cluster of several molecules or as large as a visible dust kernel. The physical and chemical properties of particulate matter are extremely varied.

Very fine particulates behave almost like a gas or vapor: they are subject to Brownian motion, follow fluid streamlines, and are capable of coagulation and condensation. Larger particulates have more of

Paul F. Fennelly is a staff scientist with the GCA/Technology Division, an environmental consulting and research firm. His current activities deal mainly with assessing the environmental impact of new energy systems and investigating the role of particulates in urban atmospheres. Dr. Fennelly received his Ph.D. in physical chemistry from Brandeis University in 1972. As a postdoctoral fellow at the Center for Research in Experimental Space Science at York University in Toronto, he was involved in a research program investigating the reactions of ions and radicals with atmospheric gases. He later joined AeroChem Research Laboratories in Princeton, N.J., where he participated in the development of techniques for monitoring atmospheric pollutants. This paper is based on a project supported by the Environmental Protection Agency; however, it should not be taken to represent the official agency view of any of the problems discussed. Address: GCA/Technology Division, Burlington Rd., Bedford, MA 01730.

Figure 1. Airborne particulates are a prime constituent of urban smog. The photograph shows an atmospheric inversion over Hartford, Connecticut, viewed from the Talcott Mountain Science Center in Avon, about 6 miles west of the city. Automobile and industrial emissions have become trapped beneath the overlying warmer air, and the murky layer containing high concentrations of sulfates, nitrates, and hydrocarbons persists until a new front restores atmospheric mixing. The plume of steam, issuing from an East Hartford industrial plant, was energetic enough to pierce the inversion but dissipated in 3–4 minutes. The horizon, barely visible as a faint gray band above the darker gray of the inversion, is about 18 miles away. The photograph was taken at about 8 a.m. on a cold winter day; by noon, the city was no longer visible through the smog. (Photo courtesy of G. C. Atamian, Talcott Mountain Science Center, Avon, CT.)

the characteristics of solid matter: they are strongly influenced by gravity and seldom coalesce or condense. The chemical behavior of particulates is determined either by the composition of the particles themselves or by the gases adsorbed by the surfaces of the particles. In some cases, the combination of particle and adsorbed gas produces a synergistic chemical effect more powerful than that of the individual components.

Particulates are usually characterized as primary or secondary. Primary particulates—usually 1 to 20 μm in size—are those injected directly into the atmosphere by chemical or physical processes. Secondary particulates are produced as a result of chemical reactions that take place in the atmosphere. They are relatively smaller and can be generally classified chemically as sulfates, nitrates, and hydrocarbons. Not surprisingly, the heaviest concentrations of airborne particulates are usually located along the coastlines and in the heavily industrialized mid-central states. However, as shown in Figure 2, air quality standards are exceeded fairly uniformly across the nation, even in rural areas—despite a general national decline in the total suspended particulate levels during the period 1970–73, when the legal controls were first put into effect. The evidence of violations even in nonindustrial rural areas is being used in many quarters to argue for a reinvestigation of the standards.

Primary particulates

Primary particulates, produced directly by the physical or chemical processes peculiar to a specified emitter, come from extremely varied sources, ranging from those in the industrial sector, such as gravel crushers and blast furnaces, to those in nature, such as forest fires and ocean spray. The chemical compositions, of course, vary with the type of source.

Particle size can be expressed either in terms of the physical or geometric diameter or in terms of an equivalent diameter pertaining to an optical, electrical, or aerodynamic property of the particle. The technique used to measure individual particles often determines which category of size is reported. For example, sieves, which are effective only in sizing particles larger than about 5 μm, measure the geometric or physical diameter. Impactors, usually effective in sizing particles in the range 0.1–5.0 μm, determine the aerodynamic diameter—a measure related to the density of the

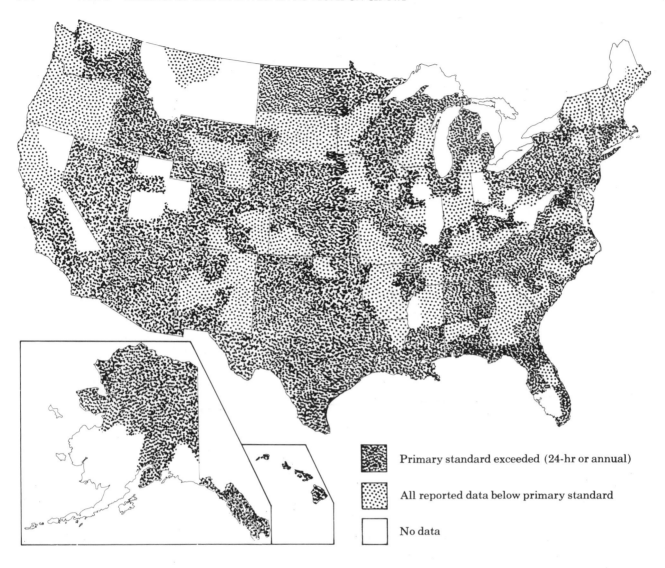

Primary standard exceeded (24-hr or annual)

All reported data below primary standard

No data

Figure 2. Although the legal controls on air pollution instituted nationwide in the period 1970–73 brought about a general decline in the level of suspended particulates, the map, based on measurements taken in 1973, shows that air quality standards continue to be exceeded in urban and rural areas alike throughout the U.S. Frequently, violations of the standards result from the presence of particulates formed by uncontrollable natural processes such as windblown dust. (Adapted from ref. *61*.)

particle and its behavior in a flowing air stream. In an impactor, the airstream containing the particulates is passed through a series of orifices and collection plates which are separated by successively smaller distances. Particles are selectively deposited on the collection plate on the basis of their inertia in the gas stream. Particles on the order of 0.1 μm can be monitored with light-scattering techniques that determine the size peculiar to the optical properties of the particle.

In compiling data on particulates, the use of standard sizing and sampling methods is very important. At present, we lack standardized techniques, and, as a result, much of the available data on particulate size is

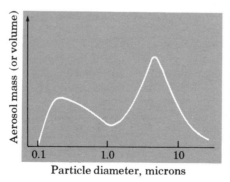

Figure 3. Recent experiments indicate that atmospheric particulates show a bimodal size distribution (represented schematically in the graph) which reflects their different formation mechanisms. Larger, primary particulates are formed by a variety of physical and chemical means; they include soil dust and solid industrial emissions released directly into the air. The smaller, secondary particulates are the product of chemical reactions taking place in the atmosphere.

difficult to assess. Until a few years ago, data compiled by traditional aerosol sizing techniques indicated that atmospheric aerosols had a unimodal size distribution. Recent data acquired with more sensitive aerosol sizing instrumentation indicate a bimodal size distribution, as shown schematically in Figure 3. Since different techniques were used it was unclear whether the observed distribution was an atmospheric effect or an instrumental artifact. The validity of the conflicting results was recently debated (*1, 2, 3, 4*), and the consensus was that smog does have a bimodal size distribution with respect to particle volume and mass and that the older sizing techniques were inadequate to detect it.

Figure 4. A wide variety of types and sizes of primary particulates are found in the aerosols of both urban and rural areas. The sizes of the species represented here are based on the geometric diameter of equivalent spheres, measured in microns. (Adapted from ref. *62*.)

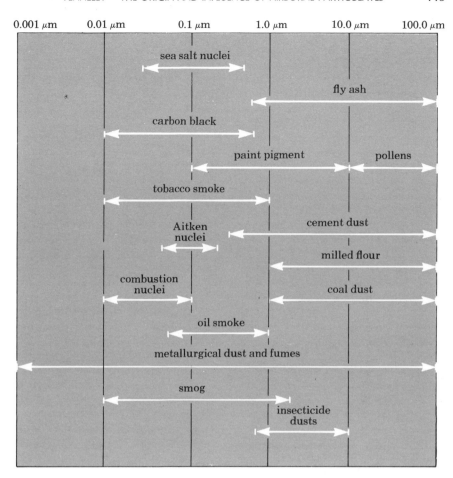

Other evidence supports this finding, indicating that most particles larger than 2.0 μm are primary particulates and most of those smaller than 1.0 μm are secondary particulates. Distinctions of this sort can have significant implications for particulate control technology. For example, the cost of removing particulates from gas streams increases drastically as the size of the particles to be removed decreases. If indeed most of the particles in the atmosphere that are smaller than 1.0 μm are secondary particulates, then the development of a technology for removing ultrafine particles from stack gases may be a waste of time and money. The effort might better be devoted to controlling the gaseous precursors of secondary particulates. Which suspected culprits air pollution control technology should try to control (and to what degree) remains a question. But until standard sizing and sampling methods can be established, it is imperative that more comparisons of data acquired with different sizing techniques be conducted.

The size ranges of some common particles are summarized in Figure 4, and Table 1 offers data on the relative contributions of various sources to the particulate loading of the atmosphere. The data in Table 1 involve gross averages and are probably reliable only to within an order of magnitude. For the most part any data based on particulate mass loading must be cautiously assessed because there is no direct correlation between the mass of particulates in the air and their effect on air quality in general. Natural dust, although it constitutes almost half the total mass of particulate matter injected into the atmosphere, has a relatively small impact. Because dust particles are generally larger than 10 μm and are fairly dense, they quickly settle out of the atmosphere as a result of gravity-induced sedimentation.

Dustfall is a nuisance, but it is usu-

Table 1. Significant sources of man-made particulate pollution in the United States

Source	Emissions (millions of tons/yr)	Total
Natural dusts		63
Forest fires[a]		56.3
Wildfire	37	
Controlled fire		
Slash burning	6	
Accumulated litter	11	
Agricultural burning	2.4	
Transportation		1.2
Motor vehicles		
Gasoline	0.420	
Diesel	0.260	
Aircraft	0.030	
Railroads	0.220	
Water transport	0.150	
Nonhighway use		
Agriculture	0.079	
Commercial	0.012	
Construction	0.003	
Other	0.026	
Incineration		0.931
Municipal incineration	0.098	
On-site incineration	0.185	
Wigwam burners (excluding forest products disposal)	0.035	
Open dump	0.613	
Other minor sources		1.284
Rubber from tires	0.300	
Cigarette smoke	0.230	
Aerosols from spray cans	0.390	
Ocean salt spray	0.340	
Total		122.715

SOURCE: Ref. *6*.
[a] More recent estimates indicate that the emission rates for forest fires are as much as a factor of 5 lower than those listed here.

ally a local condition, although some instances of long-range transport of dust have been reported. High concentrations of airborne particulates observed in the Caribbean have been correlated with severe dust storms in the Sahara (5). Long-range transport of soil dust, however, is usually limited to areas with special meteorological conditions, such as the prevailing trade winds of the Caribbean. The long-range transport of secondary particulates or very fine combustion nuclei, however, may be more common.

Transportation systems contribute only about 1% of the total mass of particulate matter, yet their impact is far more severe than that of natural dust, because most particles produced by motor vehicles are very small. Approximately 60–80% of the particulate mass in automobile exhaust is composed of particles having aerodynamic diameters smaller than 2.0 μm (7). The distribution of these smaller particles depends on atmospheric turbulence and wind conditions, and they can be transported over long distances. There is evidence that fine particulates produced in urban areas can be transported to rural areas several hundred miles away; in fact, lead particles presumably stemming from emissions in North America have been detected in Greenland's glacial ice. Further, these smaller particles are generally more hazardous to human health than their larger counterparts, both because they are more effectively embedded in the human lung and because they are typically emitted at ground level in places where population density is high.

Particles ranging from 10 to 100 μm in diameter tend to have characteristics in common with local soil conditions or effluents from local industries. In maritime areas, airborne sea salt is in this size range. Industries using grinding systems, such as grain elevators, feed mills, cement factories, and ore smelters also produce particles of this size. Friedlander (9) has recently discussed techniques for estimating the percentage contributions of various sources to the total atmospheric aerosol, using certain elements as tracers for specific pollution sources. The main requirement for a tracer element is that it not be present in large amounts in the emissions from other sources (e.g. Pb, Br for automobiles, Ca for cement, V for fuel oil). These techniques have been used with fair success in studies of California aerosols.

It is very difficult to create particles smaller than a few microns in diameter by pulverizing or grinding larger particles. The limiting factor is the large amount of energy required to provide the additional surface energy that accompanies an increase in the number of smaller particles. Thus small primary particulates (i.e. those smaller than a few microns) are produced almost exclusively in high-energy systems such as combustion engines and metal-processing furnaces.

Combustion produces particulates having very diverse chemical compositions as well as a wide range of particle sizes (0.01–10 μm). Particles are created in a combustion system in a number of ways: (1) mechanical processes may reduce either fuel or ash to particles on the order of several microns in diameter; (2) partial combustion of fossil fuels may lead to the formation of soot, with particle sizes from 0.01–1.0 μm; (3) if the fuel itself is an aerosol, a very fine ash may escape during combustion, resulting in particles on the order of 0.1–1.0 μm; (4) fuel material can vaporize and subsequently condense to form particles ranging in size from 0.1–1.0 μm; and (5) the available high energy can ionize some species and, through electrostatic interactions with other gases, the ions can generate particles of very small size, ranging from unstable clusters of a few molecules to condensation nuclei on the order of 0.01 μm.

A comprehensive summary of the chemical composition and size distribution of particles from potential sources of primary particulates has not yet been compiled. Resolution of particle sizes smaller than several microns is not included in most of the available data because early control strategies were based primarily on mass loading and ignored the potential hazards of lightweight fine particles. The extent of these hazards has only recently gained attention (10).

Another crucial lack in much of the existing data base that has just begun to be investigated is the chemical composition of particulate matter as a function of size. Recent studies (11, 12) of the chemical composition of fly ash as a function of particle size have found that toxic elements such as lead, manganese, cadmium, thallium, chromium, arsenic, nickel, and sulfur increase markedly in concentration with decreasing particle size. The mechanism controlling this selective concentration is not yet under-

Table 2. Global emission of all particulates

	Millions of tons/yr	Total
Man-made		296
Particles	92	
Gas-particle conversion		
SO_2	147	
NO_x	30	
Photochemical (hydrocarbons)	27	
Natural		2,312
Soil dust	200	
Gas-particle conversion		
H_2S	204	
NO_x	432	
NH_3	269	
Photochemical (terpenes, etc.[a])	200	
Volcanic dust	4	
Forest fires	3	
Sea salt	1,000	
Total		2,608

SOURCE: Ref. 19.

[a] Considerably higher estimates have been made for the natural production of aerosol by photochemical polymerization of terpene vapors, but the amount of experimental data available is inadequate to permit certainty.

stood, nor is it yet clear that this is a general phenomenon, since these studies were performed on stack gases from only a few power plants. Information about chemical concentrations could have important environmental consequences, for smaller particles have longer atmospheric residence times and are very effectively deposited in human lungs. Similar studies should be extended to other sources of primary particulates.

Secondary particulates

Secondary particulates, the products of chemical reactions occurring in the atmosphere, can initiate in the gas phase or as a result of reactions between gases and already existing particles. They are a major source of the ubiquitous Aitken nuclei, or solid condensation centers, that are essential for most of the condensation processes that take place in the atmosphere. They are also a prime component of urban smog.

Secondary particulates range in size from molecular clusters having diameters on the order of 0.005 μm to particles with diameters as large as several microns. Field studies of several urban aerosols (13, 14, 15) have shown that the highest concentrations of secondary particulates are usually in the range 0.01–1.0 μm. The data also indicate that within this range are two rather distinct groups, one having mean diameters less than 0.05 μm and the other having diameters from 0.05–1.0 μm. The smaller particles result directly from photochemical reactions, and the larger particles in turn are produced by the coagulation or condensation of the photochemically generated particles. The concentration of particles in both groups varies directly with intensity of sunlight and concentration of ozone.

The principal factors governing distribution by size are the respective rates of particulate formation and removal. The smallest particles, which are created constantly during the daylight hours, have lifetimes on the order of several minutes before they coagulate with larger particles (0.01–0.1 μm), which then persist up to about 12 hours before they eventually coagulate or con-

dense to form even larger particles. The overall life cycle of secondary particulates is difficult to determine; estimates range from one week to 40 days (18). In the end, the particles are either washed out of the atmosphere by rain, or they coagulate with larger primary particles that eventually settle out of the atmosphere.

The main ingredients in the formation of secondary particulates are sunlight (and to some extent cosmic rays and terrestrial radiation) and chemicals such as sulfur dioxide, ammonia, nitric oxide, water, and hydrocarbons, which enter the atmosphere from both natural and man-made sources (see Table 2). Robinson and Robbins (19) estimate that about 8% of the total global aerosol is composed of anthropogenic secondary particulates, from sources such as combustion systems, vehicle emissions, and industrial processes, and about 56% is composed of secondary particulates from natural sources, including the oceans, volcanoes and geysers, and the vapor from trees, plants, and decaying organic matter.

Mechanisms of formation

Found in both urban and rural areas, secondary particulates are in general composed of three types of chemical compounds: sulfates, hydrocarbons, and nitrates.

Sulfates. Sulfates are ubiquitous. A large fraction of the background global aerosol is ammonium sulfate $(NH_4)_2SO_4$ (20). Sulfuric acid (H_2SO_4), which along with sulfate salts such as $PbSO_4$ is found in most urban aerosols, results from the reaction of SO_3 and water. The sulfate salts, in turn, derive from the reactions of compounds such as ammonia or metallic oxides with sulfuric acid droplets. These reactions are exceedingly fast; Cadle and Robbins (21) estimate that, in the case of ammonia and sulfuric acid, reaction occurs at every collision between an acid droplet and a gas molecule. Because of the speed of the reactions, the key step in sulfate formation is the oxidation of SO_2 to form the SO_3 needed to produce sulfuric acid.

There are four basic mechanisms for the oxidation reaction that

creates SO_3: reaction with oxygen atoms, direct photochemical oxidation, catalytic oxidation, and free radical reactions. The relative importance of each mechanism in the overall atmospheric sulfur cycle is still a matter of heated debate.

SO_3 is produced by combining SO_2 with oxygen atoms in the three-body reaction

$$SO_2 + O + M \rightarrow SO_3 + M$$

where M is an inert gas such as nitrogen or argon. Recent estimates of the rate of this reaction range from 0.02 to 1.2% per hour (22), which is consistent with the observed rates of disappearance of SO_2 in clean air (about 0.1 to 0.6% per hour) (23, 24, 25).

In the lower atmosphere, the primary source of oxygen atoms for the reaction is probably NO_2, which photodissociates with practically unit efficiency when irradiated with light in the region 300–360 nanometers (26). Sunlight in this wavelength region is able to penetrate the lower atmosphere. Recent studies by Bricard et al. (27) have confirmed that NO_2 in concentrations as low as 0.5 parts per million significantly increases the oxidation of SO_2 in air when the gas mixtures are irradiated with light at wavelengths longer than 300 nanometers.

The validity of this mechanism is still in question. We do not know whether sufficient concentrations of oxygen atoms are available for reaction in the troposphere. Light capable of dissociating O_2, a potentially vast source of oxygen atoms, usually does not reach the troposphere, and the minute quantities of oxygen atoms that may be produced there can react quickly with abundant gases such as N_2 and CO; hence they should be essentially unavailable for reaction with SO_2. Other potential sources such as NO_2 may not be large enough to provide significant concentrations of oxygen atoms. At higher altitudes, however, this mechanism is almost certainly important. In the boundary between the stratosphere and the troposphere, for example, there is a significant sulfate aerosol (28), which probably results from reactions initiated by photodissociated oxygen atoms diffusing from the stratosphere.

The basic mechanism for photochemical oxidation has been postulated by a number of workers, among the first of whom were Dainton and Ivin (29):

$$SO_2 + light \rightarrow SO_2^*$$

$$SO_2^* + O_2 + M \rightarrow SO_4 + M$$

$$SO_4 + SO_2 \rightarrow 2SO_3$$

$$SO_4 + O_2 \rightarrow SO_3 + O_3$$

The principal reactive excited state of SO_2 is believed to be the long-lived triplet state (30), which is formed predominantly by intersystem crossing from the singlet state produced by absorption of light in the wavelength region 240–320 nanometers. Some triplet SO_2 results from direct absorption of sunlight (at 340–400 nanometers), but this process occurs infrequently and is probably unimportant in atmospheric photochemistry.

Experiments by Gerhart and Johnstone (31) support the validity of this mechanism, but Bricard et al. (27) suggest that the results could be explained equally well by the presence of impurities in the gas samples. There are important potential flaws in the theory: Cadle (32) has pointed out that, although reactions like the second one listed above are exceptionally fast, the quantum yields for the production of electronically excited SO_2 may be too small to be of importance to atmospheric chemistry. Further, rate coefficients for the third and fourth reactions listed above have yet to be measured and may be very slow. In fact, the species SO_4 has never been conclusively observed.

In the catalytic oxidation mechanism, SO_2 is oxidized either in aqueous droplets in the reactions

$$SO_2 + H_2O \rightarrow H_2SO_3$$

$$2H_2SO_3 + O_2 \rightarrow 2H_2SO_4$$

or on the surface of metal oxide particles such as PbO, Fe_2O_3, CaO, and Al_2O_3 (22). Iron oxide is known to be particularly effective as a catalyst both in the solid phase and in solution (22, 23), in some cases en-

hancing oxidation rates by several orders of magnitude. The catalytic effect is strong enough to produce significant oxidation of SO_2 even in the dark, demonstrating a potential mechanism for significant aerosol production at night.

A major uncertainty in these catalytic mechanisms results from our ignorance about what happens at a microscopic scale on the surface of a particle. In addition, a question remains as to whether laboratory studies have adequately duplicated atmospheric conditions.

Atmospheric sulfate formation is also enhanced by the presence of free radicals, which are often formed in mixtures of hydrocarbons and ozone or other oxidants. Ordinarily, the reaction of ozone with SO_2 takes place slowly, but in the presence of hydrocarbons it is considerably accelerated (34). Laboratory simulations of urban smog reactions have shown that SO_2 can also be oxidized rapidly in mixtures of NO_x and hydrocarbons. For certain mixtures (e.g. SO_2, 50 parts per billion; NO, 30 ppb; and cis-pentane, 100 ppb), SO_2 oxidation rates as high as 10% per hour have been reported (35). Evidence indicates that the aerosol resulting from olefin/NO_x/SO_2 mixtures is predominantly H_2SO_4 droplets for hydrocarbons with less than 5 carbon atoms; for larger hydrocarbons, carbonaceous matter also appears in the aerosol (36).

The hydrocarbon/NO_x/SO_2 mixtures can be described as synergistic systems—no two of the components react effectively, but the presence of all three provides a powerful oxidizing system. These mixtures are suspected to be major sources of sulfate particulates, but they are still the least well understood of all the oxidation mechanisms. Undoubtedly they involve the formation of intermediate, highly reactive free radical species, but which ones is still unclear. Recently, Davis and his coworkers (37, 38) have pointed to hydroxyl radicals as significant atmospheric oxidizing agents. With respect to SO_2, the proposed oxidation reaction is

$$OH + SO_2 + M \rightarrow HSO_3 + M$$

Once HSO_3 is formed, a series of re-

actions can take place with species such as O_2, NO, or hydrocarbons that ultimately results in the formation of H_2SO_4 or organic sulfates.

Of course, the relative importance of sulfate-forming mechanisms varies according to local conditions. In urban areas with relatively high concentrations of NO_2, oxidation by oxygen atoms or oxygen-containing free radicals could be important. In the plumes from large smokestacks with heavy particulate loadings and high relative humidity, heterogeneous catalytic oxidation probably predominates. In rural areas with relatively clean air, photochemical oxidation of SO_2 may be the primary mechanism of sulfate formation.

Hydrocarbons. The second major constituent of particulates is hydrocarbons, which react with oxidants (e.g. NO_2, O_3) in the atmosphere to produce peroxide radicals. Through a series of chain reactions, these radicals eventually form large organic molecules which condense to form droplets or solid particles. The mechanism is as follows:

$$NO_2 + light \longrightarrow NO + O$$

$$O + O_2 + M \longrightarrow O_3 + M$$

$$O_3 + olefins \longrightarrow RO\cdot + O_2$$

$$RO\cdot + olefins \longrightarrow R\text{-}R'O \ or \ R\overset{O}{\underset{\triangle}{-}}R$$

$RO\cdot$ represents the hydrocarbon peroxide radicals which are readily formed by ozone attack at the carbon-carbon double bond position in olefinic hydrocarbons. Once formed, organic molecules represented by R–R'O are also susceptible to photochemical excitation and further reaction. Estimates indicate that about 5% of the organic vapor in photochemical smog reacts to form particulates.

Primary sources of hydrocarbons in urban areas are automobile exhaust, power-generating stations, and industrial effluents. In rural areas, a rich source is the vapor produced by various plants, which often contains a class of compounds known as terpenes. One very common terpene, α-pinene, is found in oils of more than 400 plants. Many terpenes are highly unsaturated, and in some cases they can even be oxidized in the dark by reaction with O_2. Rasmussen and Went (39) have estimated that the worldwide emission of organic vapor is 4.4×10^8 tons per year. As mentioned earlier, the particulates formed by oxidation of these organic vapors are often the cause of the haze observed in rural areas (40).

Nitrates. The key intermediary in nitrate formation is nitric acid (HNO_3) in either vapor or droplet form. Nitric acid can be produced in the atmosphere through the following series of reactions:

$$NO_2 + O_3 \rightarrow NO_3 + O_2$$

$$NO_2 + NO_3 \rightarrow N_2O_5$$

$$N_2O_5 + H_2O \rightarrow 2HNO_3$$

Nitric acid also results directly from reactions involving hydroxyl radicals:

$$NO_2 + OH + M \rightarrow HNO_3 + M$$

After the formation of nitric acid, nitrates are created by its reaction with gaseous or solid species. For example:

$$NH_3 + HNO_3 \rightarrow NH_4NO_3$$

Ammonia is also known to form NH_4NO_3 directly in reactions with ozone (41). The mechanism is not yet understood, but it presumably involves the formation of HNO_3 in some intermediate step.

One further mechanism for the formation of nitrate particles is the adsorption of nitric acid on the metal-bearing particulates often found in urban smog. The net result is compounds such as $Pb(NO_3)_2$. Nitrate particulates are also frequently found in coastal cities and are thought to result from reactions between gaseous species and particles from ocean spray:

$$NaCl + NHO_3 \rightarrow NaNO_3 + HCl$$

Deleterious effects of particulates

Some of the ill effects of particulate matter on human health are obvious and annoying, as anyone with an allergy or hay fever will confirm. At times, however, the effects of particulates in the air can be catastrophic. Usually the most serious threats to health result from a com-

Table 3. Estimated annual health costs of air pollution, reduction in incidences of particular diseases attainable by 50% reduction in air pollution, and the resulting estimated savings (1970)

Disease	Lost income and medical expenditures (millions)	Reduction in incidences	Savings (millions)
Respiratory	$4,887	25%	$1,222
Cardiovascular	4,680	10%	468
Cancer	2,600	15%	390
Lung cancer	132	25%	33
Total savings			~$2,080

SOURCE: Ref. 49.

bination of heavy concentrations of particulate matter from industrial or man-made sources with weather conditions that prevent adequate atmospheric mixing.

Atmospheric inversion is the cause of most air pollution episodes. Normally, the temperature of the atmosphere decreases with increasing altitude, but during an atmospheric inversion, this situation is reversed, and the temperature of the air increases with increasing altitude. The warm air in the upper reaches of the atmosphere acts as a "lid" which prevents vertical mixing from below; hence, concentrations of pollutants can build up rapidly. Inversions can result from the mixing of warm and cold weather fronts, the interaction of sea and land breezes, or the radiation of heat from the land into the atmosphere.

It is estimated that 3,500 to 4,000 people died as a result of the severe fog of 5–9 December 1952 in London (42). The diseases for which mortality figures are quoted in the records of that incident are bronchitis, coronary diseases, myocardial degeneration, pneumonia, vascular lesions of the central nervous system, respiratory tuberculosis, and cancer of the lung. Other tragic air pollution episodes took place in the Meuse River Valley in Belgium in December 1930 (60 people died and 6,000 became seriously ill), in Donora, Pennsylvania, in 1948, and in London again in 1962.

The long-term effects of air pollution, and especially of airborne particulates, may be even more insidious. Specifying exactly which pollutants have the most significant effects on human health is difficult

and little definitive information is available; expanded research is needed. The length of time required for compiling accurate data has limited our assessments of these long-term effects, but some advances have been made. In many epidemiological studies, for example, particulates have been shown to have a significant effect. Growing evidence indicates a consistent relationship between exposure to particulates combined with SO_2 and impaired ventilatory function in children 5 to 13 years of age (43). In Japan, the incidence of diseases such as chronic bronchitis, bronchial asthma, and pulmonary edema has been linked to sulfuric acid mist and suspended dust particles (44). The U.S. National Academy of Sciences has said that exposure to particulate polycyclic organic matter can result in cancer of the skin and lungs, nonallergic contact dermatitis, hyperpigmentation of the skin, folliculitis, and acne (45). The effects of suspended sulfates on human health have also been examined by Shy and Finklea (46), who indicate that these particles contribute substantially to the aggravation of chronic respiratory disease.

Although it is not yet possible to isolate the actual disease-producing mechanisms or even to know which specific types of particulates are the main villains, there is a growing consensus that fine particulates (i.e. those smaller than several microns in diameter) are primary suspects. These species are especially troublesome because they are capable of bypassing the body's respiratory filters and penetrating deep into the lungs. More than 30% of the particles smaller than 1 μm that penetrate the pulmonary system remain there (47, 48). The ability of fine particulates to become embedded in the tissue is a function primarily of their geometry and is independent of their chemical composition. Once the particles have been deposited, however, their chemical nature is a prime determinant of their toxicity. As noted above, growing evidence suggests that poisonous elements, especially heavy metals such as lead, cadmium, vanadium, and nickel tend to be concentrated in these smallest particles, but even materials such as silicones, which for the most part are chemically inert, can cause acute physical irritation of sensitive lung tissue and lead to diseases such as silicosis. Particulates deposited in the lungs can also impair oxygen transfer, and, because they have fairly long lives in the atmosphere and are capable of adsorbing significant quantities of toxic gases such as SO_2 and HCl, they can produce severe synergistic effects if inhaled.

In an extensive review of the literature, Lave and Seskin (49) attempted the difficult task of estimating the cost of air pollution with respect to human health. They found significant correlations of illnesses such as respiratory disease, cardiovascular disease, and cancer with air pollution indices (e.g. dustfall, sulfation rate, concentration of suspended particulates, etc.). The correlations were so impressive that the authors point out that the only way to discredit the results would be to argue that the "real" cause of ill health was the presence of a third unknown agent which happens to correlate with levels of air pollution—an event that seems most unlikely. Their cost estimates (summarized in Table 3) are conservative and, of course, do not account for the inflation which has occurred since 1970. Perhaps a more relevant index of the cost of air pollution would be their estimate that a decrease of about 4.5% of all economic costs associated with morbidity and mortality could be achieved by a 50% reduction of air pollution in our major urban centers.

Curtailed visibility, a hazard to people traveling both by land and by air, is another way in which airborne particulates can endanger human safety. The effects of particulate matter on visibility have been

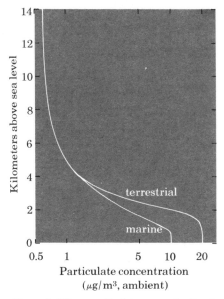

Figure 5. When particulate concentration is measured as a function of altitude, it is found that almost the total weight of particulate matter in the atmosphere is in the 2 kilometers nearest the ground. (Adapted from ref. 19.)

discussed thoroughly by several authors (50, 51). When particles scatter the light passing from the sun or some other source through the line of sight of the observer, visibility is impaired because the reduced light flux decreases the contrast between objects and their background. On the basis of a correlation of visibility with the atmospheric light-scattering coefficient and the concentration of airborne suspended particulates made by Charlson and coworkers (52), it is estimated that, in rural air where particulate concentrations are about 20 $\mu g/m^3$, visibility will be 50–60 km. In urban air where particulate concentrations are typically as high as 100 $\mu g/m^3$, visibility will be 8–10 km. The most significant reduction in visibility is generally attributed to particles in the size range 0.1–1.0 μm, which are especially effective light scatterers because their diameters are comparable to the wavelengths of light in the visible region. Small particles scatter light of shorter wavelengths (i.e. blue light) more effectively; hence, during pollution episodes, the sun often takes on a distinct red color. And, of course, the reddish hue of sunsets is also caused by the influence of atmospheric particulates.

Particulates are involved to some

extent in the transmission of odors. Sometimes they themselves are volatile, and sometimes they simply serve as a transport medium for volatile matter. Odors from restaurants, paint spraying, roofing and paving, incinerators, and open dump fires are known to be transmitted by particulates (53). If particulate matter is removed from diesel exhaust, the odor of the fumes is significantly reduced (54).

The visible effects of particulate air pollution on property are familiar to everyone: soot adsorbed on the surfaces of urban buildings is unsightly, but a hidden and more dangerous effect is the concomitant disintegration of the masonry by acids and tars adsorbed on the soot deposit. In addition, airborne particulates are capable of corroding metals and soiling and eroding metal coatings, painted surfaces, and textiles. The corrosiveness of particulates is due either to their own chemical activity or to their ability to adsorb reactive gases such as NO_2 and SO_2. The contribution of airborne particles to corrosion has been investigated by Preston and Sanyal (55). Soiling is also a problem; one study estimates that annual costs for personal property cleaning are $85 more per person in a pollution-prone city than in a clean city (56).

Particles in the atmosphere can affect the climate in several ways, but as yet the predominant effect of atmospheric particulate loading has not been determined. Field test data in both rural and urban areas indicate that concentrations of airborne particles vary widely. Mean or average concentrations must always be considered as general approximations of the atmospheric condition. Some idea of the vertical distribution of particulates can be gained from Figure 5, which indicates that almost the total weight of the aerosol is in the lowest 2 kilometers of the atmosphere. The interaction of the frictional drag of the earth's surface and the higher-level air flows occurs primarily in the first kilometer or two of the atmosphere. This region, which is called the planetary boundary layer, governs most of the transport and deposition of particulates (57) through the surface winds which predominate here.

Particles both absorb and scatter sunlight, thus reducing the amount of radiation reaching the surface of the earth and lowering surface temperatures. But particles can also absorb energy and reradiate it in the infrared spectrum, creating a "greenhouse effect" (in a greenhouse, sunlight transmitted by the glass is absorbed by the plants within; when some of the absorbed energy is reradiated in the infrared part of the spectrum, this radiation with longer wavelengths is not effectively transmitted by the glass, and the heat is trapped). Whether the presence of particulates results in a net heat loss or gain depends in a complicated way on the optical properties of the particles, the temperature structure of the atmosphere, and the degree of reflectivity, or albedo, of the surface below.

Computational models have been developed to assess the influence of the particulate greenhouse effect. In general, they indicate that the warming effects of particulates do not compensate for the surface cooling caused by attenuation of solar radiation (58). However, these averaged global models are not sufficiently sensitive to indicate significant modifications in local climate. For example, particulates originating in volcanic activity often alter local atmospheric temperature: the Mount Agung eruption in Indonesia in 1963 raised the temperature in the stratosphere by about 6 to 7°C (59). The effect of Agung was observed for about 15 degrees of latitude north and south of the source at stations around the world. Many cities generate what is sometimes called the "heat island" effect, wherein the air directly over the city is 3 to 4°C warmer than that in the surrounding countryside.

Particulates have also been implicated in altering weather patterns, but the evidence relating to this phenomenon is unclear. Both cloud formation and precipitation seem to be affected, and although both increases and decreases in their occurrence have been attributed to the injection of particulates into the atmosphere from either fires or industrial sources (60), increased particulate loadings have not so far been shown to have any systematic effect on precipitation patterns, at least in urban areas. Apparently the increase in condensation nuclei is offset by the chemical reactivity of the polluted air, which prevents the formation of ice particles.

There is evidence that particulate concentration reaches a maximum in the zone between 30 and 60° north latitude (19)—a finding that is consistent with the concentration of industrial sources in these latitudes—and that this concentration has led to a change in atmospheric electrical conductivity over the North Atlantic. Just how this might affect weather and climate patterns, however, is not well documented. The influence of particulates on both climate and weather is still poorly understood. Possibly their main impact is felt on a local, not a global, scale. Undoubtedly, our understanding of the climatic effect of airborne particulates will be subject to considerable revision as more information becomes available in the next several decades.

References

1. R. E. Lee, Jr. 1972. The size of suspended particulate matter in air. *Science* 178:567.

2. K. T. Whitby, R. B. Husar, and B. Y. H. Liu. 1972. The aerosol size distribution of Los Angeles smog. *J. Colloid Interface Science* 39:177.

3. K. T. Whitby, R. E. Charlson, W. E. Wilson, and R. K. Stevens. 1974. The size of suspended particle matter in air. *Science* 183:1098.

4. R. E. Lee, Jr. 1974. The size of suspended particle matter in air. *Science* 183:1099.

5. J. M. Prospero. 1968. Atmospheric dust studies on Barbados. *Bull. Amer. Meteorol. Soc.* 49:645.

6. Midwest Research Institute. 1971. *Particulate Pollutant System Study. Vol. I: Mass emissions.* Project No. 3326-C, Report Prepared for Air Pollution Control Office, Environmental Protection Agency.

7. P. K. Mueller, H. L. Helwig, A. E. Alcocer, W. K. Gong, and E. E. Jones. 1964. *Concentration of Fine Particles and Lead in Car Exhaust.* American Society for Testing and Materials, Special Technical Publication 352, pp. 60–73.

8. National Air Pollution Control Administration. 1969. *Air Quality Criteria for Particulate Matter*, Pub. AP-49, U.S. Dept. of Health, Education, and Welfare, Washington, D.C. pp. 23–27.

9. G. Gartrell, Jr., and S. K. Friedlander. 1975. Relating particulate pollution to sources. *Atmos. Environ.* 9:279. See also S. K. Friedlander. 1973. Chemical element balances and the identification of air pollution sources. *Environ. Sci. and Technol.* 7:235.

10. N. F. Surprenant. 1974. *R&D Program for Fine Suspended Particulates.* GCA Corp. Report, Environmental Protection Agency Contract 68-02-1316.

11. R. L. Davison, D. F. S. Natusch, J. R. Wallace, and C. A. Evans, Jr. 1974. Trace elements in fly ash. *Environ. Sci. Technol.* 8:1107.

12. R. E. Lee, Jr., and D. J. von Lehmden. 1973. Trace metal pollution in the environment. *J. Air Pollution Control Assoc.* 23:853.

13. K. T. Whitby, R. B. Husar, and B. Y. H. Liu. 1971. The aerosol size distribution of Los Angeles smog. In *Aerosols and Atmospheric Chemistry*, ed. G. M. Hidy. N.Y.: Academic Press. pp. 237–64.

14. C. E. Junge. 1963. *Air Chemistry and Radioactivity.* N.Y.: Academic Press.

15. W. E. Clark and K. T. Whitby. 1967. Concentration and size distribution measurements of atmospheric aerosols and a test of the theory of self-preserving size distributions. *J. Atmos. Sci.* 24:677.

16. C. N. Davies. 1974. Particles in the atmosphere—natural and man-made. *Atmos. Environ.* 8:1069.

17. F. W. Went. 1966. On the nature of Aitken condensation nuclei. *Tellus* 18:549.

18. R. D. Cadle. 1973. Particulate matter in the lower stratosphere. In *Chemistry of the Lower Stratosphere*, ed. S. I. Rasool. N.Y.: Plenum Press.

19. E. Robinson and R. C. Robbins. 1971. *Emission Concentration and Fate of Particulate Atmospheric Pollutants.* Final Report, SRI Project SCC-8507. Menlo Park, Calif.: Stanford Research Institute.

20. G. Megaw. 1966. *Research Progress Report, Health Physics and Medical Division*, United Kingdom Atomic Energy Association.

21. R. D. Cadle and R. C. Robbins. 1961. Kinetics of atmospheric reactions involving aerosols. *Discussions of the Faraday Society* 30:155.

22. R. S. Berry and P. A. Lehman. 1971. Aerochemistry of air pollution. *Advances in Physical Chemistry* 22:47–84.

23. P. Urone, H. Lutsep, C. M. Noyes, and J. F. Parcher. 1968. Static studies of sulfur dioxide reactions in air. *Environ. Sci. and Technol.* 2:611.

24. N. A. Renzetti and G. J. Doyle. 1960. Photochemical aerosol formation in sulfur dioxide-hydrocarbon systems. *Int. J. Air Poll.* 2:327.

25. R. A. Cox and S. A. Penkett. 1970. Photo-oxidation of SO_2 in sunlight. *Atmos. Environ.* 4:425.

26. P. A. Leighton. 1961. *Photochemistry of Air Pollution.* N.Y.: Academic Press. p. 55.

27. J. Bricard, M. Cabane, G. Madelaine, and D. Viglia. 1971. Formation and properties of neutral ultrafine particles and small ions conditioned by gaseous impurities of the air. In *Aerosols and Atmospheric Chemistry*, ed. G. M. Hidy, N.Y.: Academic Press. p. 27.

28. R. D. Cadle, A. L. Lazrus, W. H. Pollack, and J. P. Sheldovsky. 1970. In *Proceedings of the Symposium on Tropical Meteorology*, ed. C. S. Ramage. Boston, Mass.: American Meteorological Society.

29. F. S. Dainton and K. J. Ivin. 1950. The photochemical formation of sulphinic acids from sulfur dioxide and hydrocarbons. *Transactions of the Faraday Society* 46:374–82.

30. H. W. Sidebottom, C. C. Badcock, G. G. Jackson, J. G. Calvert, G. W. Reinhart, and E. K. Damon. 1972. Photo-oxidation of sulfur dioxide. *Environ. Sci. and Technol.* 6:72.

31. E. R. Gerhart and H. F. Johnstone. 1955. Photo-oxidation of SO_2 in air. *Industrial and Engineering Chem.* 47:972.

32. R. D. Cadle. 1972. Formation and chemical reactions of atmospheric particles. In *Aerosols and Atmospheric Chemistry*, ed. G. M. Hidy. N.Y. Academic Press. p. 143.

33. H. F. Johnstone and D. R. Coughanower. 1958. Absorption of sulfur dioxide from air. *Industrial and Engineering Chem.* 50:1169.

34. W. E. Wilson, A. Levy, and D. Wimmer. 1972. A study of SO_2 in photochemical smog, II. The effect of SO_2 on formation of oxidant. *J. Air Pollution Control Assoc.* 22:311.

35. R. A. Cox and S. A. Penkett. 1971. Photo-oxidation of atmospheric SO_2. *Nature* 229:486.

36. P. J. Groblicki and G. J. Nebel. 1971. The photochemical formation of aerosols in urban atmospheres. In *Chemical Reactions in Urban Atmospheres*, ed. C. S. Tuesday. N.Y.: Elsevier. p. 241.

37. D. D. Davis, G. Smith, and G. Klauber. 1974. Trace gas analysis of power plant plumes via aircraft measurement: O_3, NO_x and SO_2 chemistry. *Science* 186: 733.

38. W. Payne, L. Stief, and D. D. Davis. 1973. A kinetic study of the reaction of HO_2 with SO_2 and NO. *J. Amer. Chem. Soc.* 95:7614.

39. R. A. Rasmussen and R. W. Went. 1965. Volatile organic material of plant origin in the atmosphere. *Proc. U.S. National Academy of Sciences* 53:215.

40. F. W. Went. 1960. Blue hazes in the atmosphere. *Nature* 187(4738):641.

41. K. J. Olszyna, R. G. de Pina, M. Luria, and J. Heicklen. 1974. Kinetics of particle growth, IV. NH_4NO_3 from the NH_3-O_3 reaction revisited. *J. Aerosol Sci.* 5: 421.

42. A. J. Lindsey. 1971. Air pollution and health. *Chemistry and Industry* (London) 14:378.

43. C. M. Shy, V. Hasselblad, R. M. Burton, C. J. Nelson, and A. A. Cohen. 1972. *Results of Studies in Cincinnati, Chattanooga, and New York*. American Medical Association Conference, Chicago, Ill.

44. Pollution Damages to Human Health and Countermeasures. n.d. In *Quality of the Environment in Japan*. Japan Environmental Agency. pp. 105–30.

45. Committee on Biologic Effects of Atmospheric Pollutants, National Academy of Sciences. 1972. *Particulate Polycyclic Organic Matter*. Environmental Protection Agency Contract CPA 70-42.

46. C. M. Shy and J. F. Finklea. 1973. Air pollution affects community health. *Environ. Sci. and Technol.* 7:204.

47. National Air Pollution Control Board. 1969. *Air Quality Criteria for Particulate Matter*. U.S. Dept. of Health, Education, and Welfare. Pub. No. AP-49. Washington, D.C.

48. Task Group on Lung Dynamics. 1966. Deposition and retention models for internal dosimetry of the human respiratory tract. *Health Physics* 12:173.

49. L. B. Lave and E. P. Seskin. 1970. Air pollution and human health. *Science* 169:723.

50. E. Robinson. 1969. Effects of air pollution on visibility. In *Air Pollution*, ed. A. C. Stern. N.Y.: Academic Press.

51. W. E. K. Middleton. 1952. *Vision Through the Atmosphere*. Toronto: Univ. of Toronto Press.

52. R. J. Charlson. 1968. Atmospheric aerosol research at the University of Washington. *J. Air Pollution Control Assoc.* 18:652.

53. National Air Pollution Control Board. 1969. *Air Quality Criteria for Particulate Matter*. U.S. Dept. of Health, Education, and Welfare. Pub. No. AP-49. p. 107.

54. R. H. Linnel and W. E. Scott. 1962. Diesel composition and odor studies. *J. Air Pollution Control Assoc.* 12:510.

55. R. Preston and B. Sanyal. 1956. Atmospheric corrosion by nuclei. *J. Appl. Chem.* 6:28.

56. I. Michelson and B. Tourin. 1966. *Comparative Method for Studying Costs of Air Pollution*. Public Health Report 81(6):505.

57. J. H. Seinfeld. 1975. *Air Pollution: Physical and Chemical Fundamentals*. N.Y.: McGraw-Hill. p. 10.

58. *Inadvertent Climate Modification*. 1971. Report of Man's Impact on Climate (SMIC). Cambridge, Mass.: MIT Press. p. 22.

59. *Man's Impact on the Global Environment—Assessment and Recommendations for Action*. 1970. Report of the Study of Critical Environmental Problems. Cambridge, Mass.: MIT Press. p. 104.

60. G. M. Hidy, W. Green, and A. Alkeyweeny. 1972. Inadvertent weather modification and Los Angeles smog. In *Aerosols and Atmospheric Chemistry*, ed. G. M. Hidy. N.Y.: Academic Press. p. 339.

61. *Monitoring and Air Quality Trends Report, 1973*. 1974. U.S. Environmental Protection Agency, Doc. No. EPA-450/1-74-007.

62. *CRC Handbook of Chemistry and Physics*. 1971. 51st ed. Cleveland: Chemical Rubber Company. p. F-199.

"It certainly becomes uncomfortable when the pollutants are up to 990,000 parts per million."

Owen B. Toon
James B. Pollack

Atmospheric Aerosols and Climate

Small particles in the Earth's atmosphere interact with visible and infrared light, altering the radiation balance and the climate

Small liquid and solid particles are ubiquitous in the atmosphere of Earth and the planets. With each breath an average terrestrial urban dweller may inhale 10^8 particles, or aerosols, composed of such diverse substances as sulfuric acid, sulfates of ammonia, nitrates, soil, and hundreds of organic compounds, some of which are carcinogenic. One out of four photons of visible light vertically traversing a typical urban atmosphere encounters an aerosol before reaching the ground, while a photon horizontally crossing a city may travel only about 10 km before intercepting a particle. During smoggy conditions, light is affected even more strongly by the particles. Pronounced optical effects due to aerosols are also noticeable outside the cities. Photochemically produced aerosols spread out over wide regions around cities; aerosols raised during desert dust storms are transported thousands of kilometers; and, after large volcanic eruptions, aerosols in the stratosphere can blanket the Earth (Fig. 1).

Atmospheric particles composed mainly of water are usually called cloud particles, despite their basic similarity to particles composed of other substances, which are classified

Owen B. Toon and James B. Pollack are research scientists at the National Aeronautics and Space Administration's Ames Research Center. Dr. Toon received his Ph.D. from Cornell University, and Dr. Pollack, from Harvard. Both scientists have an avid interest in studies of planetary atmospheres as well as in studies of aerosols and the Earth's climate. Dr. Pollack recently received the AIAA Space Sciences Award for his work on the effects of aerosols on climate. Address: Space Science Division, Ames Research Center, NASA, Moffett Field, CA 94035.

as aerosols. Clouds are affected by aerosols because every cloud drop in the sky forms around a particle. Aerosols can alter the optical properties of clouds by changing the cloud droplet composition, they can control precipitation by modifying the number and size of water drops, and they can alter cloud chemistry and lead to acid rainfall. Acid rainfall is a significant hazard to vegetation, to life in lakes and streams, and to exposed buildings and statues.

Other planets are even more influenced by aerosols than is Earth. On Mars, global dust storms dirty the atmosphere. Soil dust is the radiatively dominant material in the Martian atmosphere and is intimately tied to Martian climate change. On Venus, a dense photochemical smog, similar to the acid hazes of many terrestrial cities, blankets the entire planet and plays a significant role in Venusian meteorology. Aerosol clouds of various compositions also dominate the atmospheres of the outer planets Jupiter, Saturn, Titan, Uranus, and Neptune.

Aerosol studies for Earth and the planets are currently being carried out by a large number of scientists working in diverse fields. A particularly interesting controversy, which we will focus on in this paper, concerns the impact of terrestrial aerosols on the Earth's solar and infrared radiation budget and climate. Two questions are being debated: Do aerosols warm or cool the Earth? Is the effect climatologically significant?

The debate is caused partly by a lack of experimental data on the optical properties of aerosols and partly by our inability to predict accurately the

Earth's weather and climate. However, much of the confusion is due to an oversimplification of the problem that results from considering only one type of particle. Given a single type of aerosol, theorists who study the radiation budget agree reasonably well upon the magnitude of the climatic impact and on whether it should be warming or cooling. Unfortunately, real aerosols are temporally and spatially diverse. Therefore, since there is no single type of aerosol, theorists provide different solutions to the problem of the impact of aerosols on climate for different aerosols.

In this paper we will first define the optical properties that are important for characterizing the effect of aerosols on the radiation budget. Then we will discuss theoretical studies of the connection between the aerosol optical properties and the climate. After describing what is known about the optical properties of natural aerosols and their variability, we will review the available observational evidence linking aerosols and climate. Figure 2 presents a schematic summary of the climatically significant interactions between aerosols and light from the sun as well as between aerosols and infrared light from the Earth's surface and atmosphere.

Optical properties

The physical properties of aerosols, such as size, shape, refractive index, and concentration in the atmosphere, control the aerosol interaction with light according to a set of derived properties, which are known as optical properties. Three fundamental properties are the optical depth, a measure of the size and number of particles present in a given column of air; the single scattering albedo, the fraction of light intercepted and scattered by a single particle; and the asymmetry parameter, an integrated measure denoting the portion of light scattered forward in the direction of the original propagation and the portion scattered backward toward the light source. (For a detailed review see Hansen and Travis 1974.)

A light ray traversing an aerosol-laden column of air will be reduced by the effects of absorption and scattering as the exponent of the optical depth, τ, which is basically just the weighted product of the number of particles in a column of unit area and the cross-sectional area of a single particle. The weighting takes into account several optical phenomena. For example, a particle much smaller than the wavelength of light does not interact efficiently with light, so its full cross-sectional area is not counted. For visible light, a particle larger than about 0.1 μm contributes its entire cross-sectional area, or even a factor of 2 more than its area, to τ; particles of this size are very common in the atmosphere. Particles larger than about 1 μm interact most effectively with infrared radiation but are less common. The quantity τ varies in space and time from about 0.01 to 1.0 at visible wavelengths. Adequate space and time averages for τ are major observational unknowns, though τ is known for many local areas.

light, and when small quantities of these constituents are present, $\tilde{\omega}_0$ can be reduced to about 0.5. The worldwide and even the local prevalence of absorbing compounds is poorly known. Most aerosols strongly absorb infrared light, and thus $\tilde{\omega}_0$ is less than 0.1 in the infrared. The product $\tilde{\omega}_0\tau$ is the scattering optical depth: a light ray traversing an aerosol-laden column will be reduced due to scattering as the exponent of $\tilde{\omega}_0\tau$. The absorption optical depth is $(1 - \tilde{\omega}_0)\tau$.

Figure 1. The eruption of Krakatoa, in 1883, injected large quantities of volcanic gas and ash into the atmosphere. Some of the ash can be seen falling out of the volcanic plume in this lithographic reproduction of a photograph taken early in the eruption.

the asymmetry parameter, g, which varies from -1 to 1. If g were 1, all the light would be scattered into the hemisphere centered on the direction of the light beam's original propagation. If g were -1, all the light would be scattered backwards. For typical aerosols g is greater than zero. Due to the small particle size relative to infrared wavelengths, g is about 0.5 or less for most aerosols in the infrared. At visible wavelengths, g tends to be close to 0.7. Fortunately, g depends only weakly upon the particle size and composition, and thus the observational uncertainty about g is much less significant for climate than the uncertainty about $\tilde{\omega}_0$ and τ.

Aerosols are not uniformly distributed over the Earth, particularly in the lower atmosphere. At present, research into the effects of aerosols on climate is evolving from simple studies of global- and time-averaged problems toward more complicated regional, temporally varying problems. However, a basic understanding of the dependence of climate on the aerosol optical properties is most easily gained by reviewing global-average calculations (see Ramanathan and Coakley 1978 for modeling assumptions). In such calculations, atmospheric motions are generally ignored. The climate change, represented by a change in the global mean surface temperature, is found

A fraction of the light intercepted by a particle is scattered and a fraction is absorbed. The fraction scattered by any single particle is called the single scattering albedo, $\tilde{\omega}_0$. Most atmospheric particles do not strongly absorb visible light, so $\tilde{\omega}_0$ for these particles is between 0.9 and 1.0. However, there are minor aerosol constituents such as soot that do absorb visible

The light scattered by aerosols is not scattered uniformly in all directions but has a complex angular distribution that can be a function of wavelength. Rainbows and diffraction coronas, most often seen in water clouds, are due to the wavelength dependence of the angular scattering distribution. A simple integrated measure of the angular scattering is

by balancing the solar energy absorbed by the gases and aerosols in the atmosphere and by the Earth's surface against the infrared energy radiated to space by the Earth's surface and atmosphere. The balance is required in order to conserve energy.

Climatic changes, such as droughts or

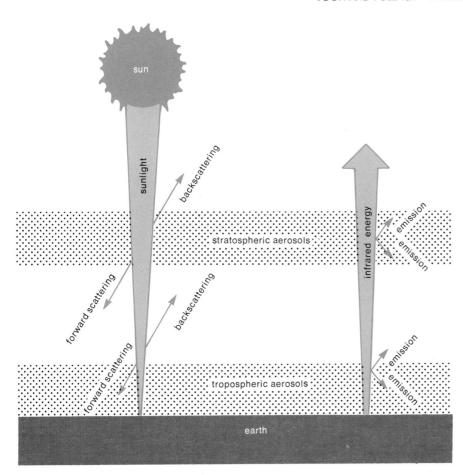

Figure 2. This schematic drawing shows the interactions between aerosols and sunlight, and aerosols and infrared light. Aerosols in the stratosphere, 20 km above the Earth's surface, absorb light from the sun and infrared light coming up from the lower atmosphere. Stratospheric aerosols have a warming effect on the surface by emitting infrared light toward the lower atmosphere, but this effect is offset by the aerosols' absorption of solar energy and backscattering of sunlight into space, which prevents it from reaching the surface. Whether the net effect of the stratospheric aerosols is to warm or cool the surface depends upon the size of the particles: since they are generally rather small, the effect is to cool the surface. Tropospheric aerosols, in the atmosphere just above the Earth's surface, cool it by backscattering sunlight, but they warm it by absorbing sunlight. Whether the net effect is warming or cooling depends upon the ratio of absorption to scattering, and this ratio depends upon the composition of the tropospheric aerosols, which is variable and poorly known.

a series of harsh winters, are usually restricted to small areas of the Earth. Often there will be compensating climatic changes from region to region: drought in one location may be nearly balanced by unusually high precipitation in another location. Periods of climatic extremes in numerous portions of the globe have been found to accompany small changes in global temperatures. We can calculate small changes in global temperature using global-average climate models, but not local temperature and precipitation changes—the quantities of real climatic importance. The local changes must simply be inferred from the global-average calculations on the basis of past climatic changes. Typical changes in the Earth's global mean surface temperature over the last thousand years, during which the climate has varied considerably, have been about 1°C, whereas the difference between the present and the ice-age mean temperatures is only about 5°C. Calculated global temperature changes of even several tenths of a degree are therefore thought to be quite significant.

Many people have studied the relationship between surface temperature and aerosol loading. Figure 3 illustrates the dependence of the surface temperature upon the tropospheric aerosol optical properties τ and $\tilde{\omega}_0$, as calculated with a relatively sophisticated global-average radiative model (Hansen et al. 1979). The surface temperature scale is zeroed to a calculation using a "standard" aerosol distribution based upon an early analysis of the available observations (Toon and Pollack 1976). The standard model has a τ of 0.125 and a $\tilde{\omega}_0$ of 0.99 at a visible wavelength of 5500 Å, while the optical depth is 0.06 at an infrared wavelength of 10 μm. Most of the aerosols in the standard model are located close to the Earth's surface and are referred to as tropospheric aerosols.

The radiation budget of the Earth is dominated by water vapor and clouds, which raise the Earth's surface temperature about 15°C above the temperature that the Earth would have without an atmosphere. Aerosols have a smaller but still significant effect. For example, Figure 3(*top left*) suggests that if there were no tropospheric aerosols, the Earth would be about 1.5°C warmer than it is. But if there were twice as many aerosols of the type found in the standard model, the Earth might be cooler than it is by 1.5°C. The reason for these differences is that the aerosols in the standard model do not absorb much solar radiation but they do scatter sunlight back into space, so less sunlight is available to warm the Earth's surface. As the aerosol concentrations increase, the surface would get progressively cooler. The aerosols are too small to interfere efficiently with infrared radiation.

The importance of the visible absorption is illustrated in Figure 3(*bottom left*). As $\tilde{\omega}_0$ decreases, the tropospheric aerosols absorb more solar energy and scatter less back to space, and their net cooling effect becomes less and less. Below $\tilde{\omega}_0$ of about 0.85, the critical albedo $\tilde{\omega}_c$, aerosols do not cool the Earth at all but absorb so much solar energy that they warm the Earth. (In this figure, the temperature scale is zeroed relative to the temperature calculated by using, as a "standard," aerosols that cool the Earth by about 1.5°C. If the temperature scale had been zeroed relative to the temperature calculated by assuming aerosols that neither warm nor cool the Earth, then the critical albedo would be at 0 of the temperature scale in Figure 3, *bottom left*).

The precise value of $\tilde{\omega}_c$ depends upon the model used. For example, several studies have shown that $\tilde{\omega}_c$ is much higher if the tropospheric aerosols are over very bright terrain such as a desert or a snow field rather than land, which has a typical albedo of 10%. This difference in $\tilde{\omega}_c$ occurs because the main climatic effect of the aerosol is to scatter sunlight back into space. However, if the ground already reflects most of the sunlight, the aerosols cannot much increase the total amount of sunlight reflected to space. Then the energy absorption by the aerosols may be greater than the scattering, causing a warming effect. The critical albedo also depends upon the asymmetry parameter, g, because if light is scattered forward, as most of it is, it still reaches the ground to warm the surface.

Figure 3(*bottom right*) illustrates the sensitivity of the surface temperature to the optical depth of aerosols at infrared wavelengths. Typical terrestrial aerosols have strong absorption bands near a wavelength of 10 μm. On the other hand, Earth's atmosphere is almost transparent near 10 μm, and consequently a large fraction of the infrared radiation escaping from the Earth to space is close to this wavelength. Aerosols prevent some of this radiation from escaping, which makes the Earth warmer through a "greenhouse" effect. As the infrared optical depth increases, the greenhouse effect warms the Earth more and more, thereby counteracting the cooling effect caused by the aerosols' scattering of solar radiation back to space. However, even in the extreme case when the infrared and visible optical depths are equal, the net effect of the tropospheric aerosols in the standard model would be to cool the Earth.

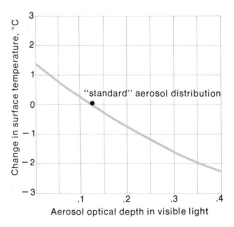

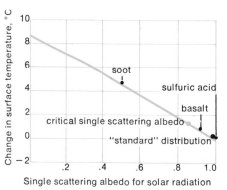

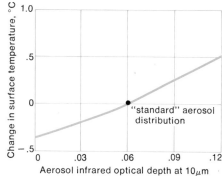

Figure 3. Surface temperature is greatly influenced by the basic optical properties of the aerosols in the troposphere. *Top left:* The surface temperature changes as the amount of aerosols varies from the amount in the "standard" aerosol distribution. *Bottom left:* The surface temperature changes as the single scattering albedo is varied from the value in the "standard" aerosol distribution; the critical albedo is the value below which the cooling effect of the aerosols would be eliminated. *Bottom right:* The surface temperature changes as the optical depth of the aerosols at infrared wavelengths, where little gaseous absorption occurs, varies from the value in the "standard" aerosol distribution. (From Hansen et al. 1979.)

The ratio of infrared to visible opacity is controlled by the size distribution of the aerosols; the standard aerosol distribution assumes a relatively great abundance of large particles, so it is not likely that the standard distribution underestimates the infrared effect.

Most aerosols are near the Earth's surface, but an important layer of particles is found about 20 km above the surface, in the stratosphere. These aerosols normally have such a small τ that they are unimportant for climate. However, after large volcanic eruptions, their τ can be as large as that of the aerosols in the troposphere. The relation of climate to the optical properties of stratospheric aerosols is slightly different than for tropospheric aerosols, as has been demonstrated by several groups (Pollack et al. 1976a, b; Hansen et al. 1978).

Figure 4 shows that the relation between the optical depth and surface temperature for the stratospheric aerosols is quite close to that given in Figure 3 for tropospheric aerosols. However, for stratospheric aerosols,

the dependence of warming and cooling upon $\tilde{\omega}_0$ and upon the variation of τ with wavelength is different than for tropospheric aerosols. The stratosphere is not strongly coupled to the Earth's surface by atmospheric dynamics, and if the stratospheric aerosols absorb some incoming solar energy, the energy would not be conducted to the surface. Indeed, it is found both observationally and theoretically that the presence of stratospheric particles simultaneously causes the surface to cool and the stratosphere to warm. The aerosols warm the stratosphere by absorbing solar and infrared energy. The solar energy which the stratospheric aerosols absorb and backscatter does not reach the surface, causing cooling there. Hence, stratospheric aerosols cool the surface for all values of $\tilde{\omega}_0$.

The size of the stratospheric aerosols is an important factor in determining whether the climate warms or cools. Small particles, as shown in Figure 4, tend to cool the surface, whereas larger ones warm the surface. For stratospheric particles whose size is less than 0.1 μm, the ratio of the infrared τ to the visible τ is 0.1. For

0.25-μm-radius particles, however, the ratio is 0.5, and for 0.5-μm-radius particles the two optical depths are nearly equal. The reason infrared opacity is more important for stratospheric aerosols is that they are at high altitude and low temperature. The larger cold particles are very effective at blocking infrared radiation coming up from the atmosphere and surface below, thereby warming the surface. Because tropospheric aerosols lie below much of the atmosphere and are almost as warm as the ground,

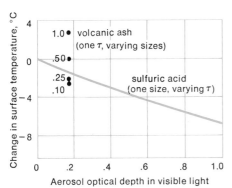

Figure 4. The surface temperature is influenced by the number of sulfuric acid particles in the stratosphere and by the size of volcanic dust particles. Large particles tend to warm the surface whereas small ones cool it; the sulfuric acid particles are usually smaller than 0.1 μm, and thus they have a cooling influence. (From Pollack et al. 1976a, b.)

they are not nearly as efficient at blocking the escape of the infrared energy.

Globally distributed stratospheric aerosols will always warm the stratosphere. However, the Earth's surface will cool if most of the aerosols are smaller than a critical size—near 0.5 μm, depending somewhat upon the aerosols' composition and the Earth's albedo. Hence τ and the size of the stratospheric aerosols are the most important parameters needed to estimate the effect of the aerosols on the climate. The major complicating factor is that, after a volcanic eruption, the aerosols evolve rapidly in size and composition as they spread out over the Earth. Although observational data are sparse, it seems that the size of the particles varies for several months: as the dust veil begins to form, the particles may be larger than the critical size.

As for the tropospheric aerosols, the Earth's surface will cool if globally distributed aerosols have a $\tilde{\omega}_0$ larger than the critical albedo (near 0.85), and the magnitude of the cooling will depend in a nearly linear manner upon τ. Hence, $\tilde{\omega}_0$ and τ are the principal variables that need to be known in order to relate tropospheric aerosols and climate. The major complication for the analysis is that aerosols are not uniformly distributed, either horizontally or vertically, throughout the troposphere, and therefore $\tilde{\omega}_0$ and τ vary widely with location. The value of the critical albedo also varies with location because it depends upon the ground albedo. In addition, once the aerosols affect the radiation budget in one area, the atmosphere responds with complex changes in the wind and clouds. Because aerosols are heterogeneous, climate theorists cannot provide a single answer to the question of whether tropospheric aerosols cool or warm the Earth: there is no single answer.

Variability of optical properties

The most important optical properties needed to determine the temperature at the Earth's surface are the stratospheric and tropospheric optical depths, the aerosol size distribution in the stratosphere, and the tropospheric aerosol single-scattering albedo at visible wavelengths. At present it is only possible to formulate a general outline of the range of these values (Toon and Pollack 1976).

The aerosols in the atmosphere form a complex mixture (Fig. 5), which can be separated into classes on the basis of the aerosols' elemental composition. Soil-derived aerosols, composed of minerals and organic debris of fairly large size (1–10 μm), comprise a large fraction (about 30%) of the global aerosol mass burden. Such aerosols spread far from the land of their origin, as evidenced by the great quantity of dust from the Sahara Desert that reaches the Caribbean (Carlson and Prospero 1972). Sea-spray aerosols, composed mainly of sea salt and marine organics, are also fairly large (1–10 μm) but constitute only a modest fraction of the total aerosol mass (10–15%). Sea salt does not spread very far inland or to very high altitude because it is rapidly removed from the atmosphere by pre-

cipitation (Junge 1972). Sulfur compounds, mostly sulfuric acid and ammonium sulfate, constitute the largest fraction (about 50%) of the total tropospheric aerosol load and are found everywhere from the continents and the oceans to the poles. Sulfuric acid is the dominant aerosol in the stratosphere. Sulfates are primarily formed within the atmosphere by complex photochemical and solution reactions and are typically rather small (0.1 μm). Many other materials are also found in the atmosphere, including graphitic soot, forest-fire debris, nitrates, and numerous organic compounds. The optical properties of these various materials are quite different.

Experimental projects to determine $\tilde{\omega}_0$ have used one of three approaches: direct measurements of $\tilde{\omega}_0$ in the atmosphere (Herman et al. 1975; Weiss et al. 1978); laboratory measurements of the absorption by collections of atmospheric aerosols and calculation of $\tilde{\omega}_0$ (Lindberg 1975; Patterson et al. 1977); or laboratory measurements of the absorption by pure materials known to compose aerosols and calculation of $\tilde{\omega}_0$ (Palmer and Williams 1975; Toon et al. 1976; Twitty and Weinman 1971). The laboratory studies have shown that materials such as sulfates and sea salt are very transparent to visible light and have $\tilde{\omega}_0$ very close to unity. Most rock particles are only moderately absorbing and have $\tilde{\omega}_0$ above 0.9, although smaller values are occasionally found. The laboratory studies combined with observations of the most common aerosol materials suggest that $\tilde{\omega}_0$ is usually well above the critical value of 0.85, and thus we would expect that aerosols will generally cool the Earth. However, in certain regions wind-borne dark aerosols derived from soil over high-albedo surfaces might cause warming. In contrast, the direct studies of $\tilde{\omega}_0$ in the troposphere find that $\tilde{\omega}_0$ is normally at or well below the critical value, with a few exceptions.

Accurate direct measurements of $\tilde{\omega}_0$ are quite difficult to make, and the available values could be wrong. Also, measurements have not been performed in enough locations to obtain a "typical" value. Most of the lowest $\tilde{\omega}_0$ values have been found in urban regions, which comprise only a small fraction of the Earth's area. For these reasons, the available direct studies may be misleading when applied to the global problem. However, it is also quite possible that the laboratory studies have overlooked minor, highly absorbing materials such as iron oxides or soot. If 10 to 20% of the aerosols were composed of soot, the value of $\tilde{\omega}_0$ would be below the critical value even if the remaining bulk of the material were completely transparent. If the absorbing particles were much smaller than the typical aerosols, an even smaller mass fraction could be very significant (Bergstrom 1973). It has been found that a small amount of soot is responsible for the low values of $\tilde{\omega}_0$ in urban areas (Rosen et al. 1978).

Very little is known about the variability of $\tilde{\omega}_0$, and much of the debate about whether aerosol pollution will lead to a warming or cooling of the Earth depends upon which compounds are being added to the atmosphere. Those who believe anthropogenic aerosols are warming the climate point to the fact that $\tilde{\omega}_0$ is lowest in urban areas, suggesting that man is lowering $\tilde{\omega}_0$ globally by emitting soot and perhaps opaque hydrocarbons. One reason for restricting soot emissions from diesel engines is to prevent the lowering of $\tilde{\omega}_0$. Many of those who believe anthropogenic aerosols are cooling the Earth blame sulfate compounds, which are known to have a large $\tilde{\omega}_0$. Human activities now probably account for about half the sulfur compounds in the atmosphere (Bach 1976). Concern about sulfate aerosols is one reason for strict emission controls on sulfur dioxide. Another group points to the large quantity of soil debris that humans release to the atmosphere by agricultural activity. Most soil debris has $\tilde{\omega}_0$ above the critical value, and thus in most areas agricultural activity may have the effect of cooling the Earth.

Since aerosol optical depth is primarily a measure of the aerosol concentration, increased levels of pollution could be detected by monitoring τ. Although it is difficult to measure $\tilde{\omega}_0$, it is quite simple to measure τ. As the solar beam passes through the atmosphere, some sunlight is absorbed by aerosols and some is scattered and ends up as diffuse sky light. The optical depth is the natural logarithm of the ratio of the sunlight

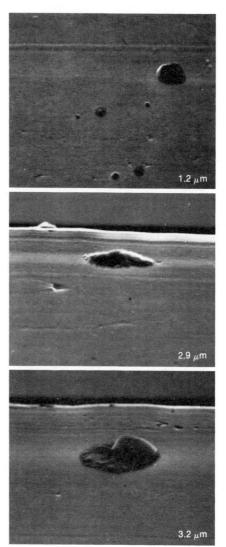

Figure 5. Stratospheric particles (*above*) are generally fluid sulfuric acid, though solid particles of diverse types are present, especially after large volcanic eruptions. These electron-microscope photographs show particles collected by N. Farlow using NASA's U2 aircraft flying near 20 km. (The number in each is the size of the largest particle.) *Below:* Tropospheric particles are extremely diverse, and their structure depends strongly upon when and where they are captured. This electron-microscope photograph shows particles obtained by D. Wood between 3 and 8 km altitude near Dallas, TX. The largest particles are about 5 μm in diameter, and x-ray diffraction studies indicate that they are soil particles; the small particles are probably sulfates.

reaching the ground in the direct solar beam to the sunlight impinging at the top of the atmosphere.

Observations of τ at visible wavelengths have been made since about 1900 at a few locations, but data from even a few places are useful for monitoring the stratospheric optical depth, because changes in the stratosphere occur on a nearly global scale. Figure 6 illustrates that changes in the stratospheric optical depth due to explosive volcanic eruptions have more than doubled the Earth's average aerosol optical depth for a year or two following each eruption. Explosive volcanic eruptions are episodic, and though many volcanoes erupted between 1870 and 1912, there has been only one large eruption since 1912. Indirect evidence (Lamb 1970) shows that there were large volcanic optical depths for the entire four centuries from 1500 to 1900, and studies of volcanic ash in polar cores reveal a 10-thousand-year period of explosive volcanic activity centered at the peak of the last ice age, twenty thousand years ago (Gow and Williamson 1971). NASA now has a satellite program designed to study the optical depths and the spread of aerosols after large volcanic eruptions.

The optical depth changes in the stratosphere following large volcanic eruptions are due partly to volcanic ash and partly to sulfuric acid particles. Unfortunately, we do not precisely know the ratio of these materials or their sizes. It is believed that ash particles larger than the critical size dominate the volcanic cloud initially but soon fall out of the atmosphere. Then sulfuric acid particles, formed from chemical reactions involving volcanic gas, dominate the period following the eruption (Toon and Pollack 1976). In order to test these ideas, NASA has an extensive aircraft sampling program in the stratosphere, and theoretical models of the aerosols have been constructed in an attempt to predict the aerosol sizes (Turco et al. 1979).

Considerable thought has been given to the possibility that human beings might alter the stratospheric aerosol optical depth by flying aircraft and rockets through the stratosphere or by adding sulfur gases to the stratosphere as industrial pollutants (Pollack et al. 1976b, c; Turco et al. 1980). At present, it does not appear that any of man's planned activities during the next several decades will cause a significant enhancement in the stratospheric optical depth, except possibly a small increase due to the release of carbonyl sulfide, an industrial pollutant (Turco et al. 1980).

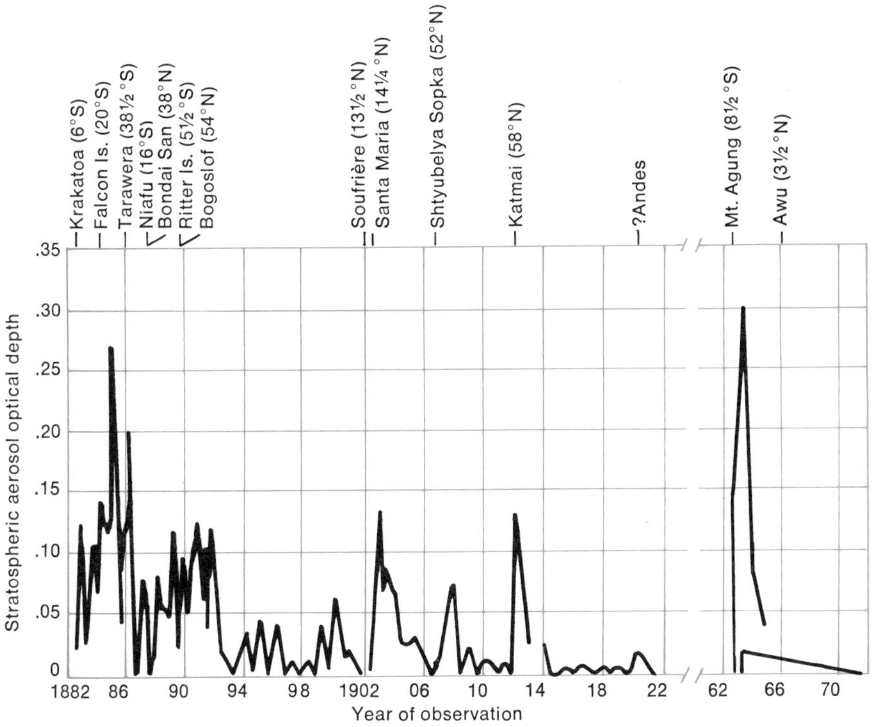

The optical depths of tropospheric aerosols vary on shorter time and space scales than do those of stratospheric aerosols. A general trend

areas, or agricultural areas, including parts of Japan, the eastern and southwestern United States, and the Soviet Union, we have evidence of large upward trends in optical depth during the past several decades (Yamamoto et al. 1971; Husar et al. 1979; Machta 1972; Trijonis 1979). Tropospheric aerosols come from local sources and are rapidly removed from the atmosphere. Hence the major changes are expected on regional but not global scales. At present, we have an incomplete picture of the regional

regions and during droughts in semi-arid regions of the world. For example, the optical depth of the dust from the Sahara Desert often reaches 0.5 over the Atlantic and sometimes exceeds 2.0 (Carlson and Caverly 1977). Also, large variations in the dust over the Atlantic may be partially caused by drought in Africa (Prospero and Nees 1977). A well-known example of drought-caused dust was the Dust Bowl era of the 1930s, which affected the plains of the south-central United States. Although much dust is naturally released from the soil, human activities have greatly increased the amount supplied to the atmosphere by altering the natural vegetation that used to prevent soil from being eroded. No observations allow us to estimate exactly how much of an impact these soil particles have on the global average optical depth, but local anthropogenic enhancements are well known.

Figure 6. Observations of the change in stratospheric aerosol optical depth at visible wavelength over the past century show that large volcanic eruptions create optical depths for a year or two that are as large as those of tropospheric aerosols. The optical depths vary over the Earth. For example, the data after the 1963 eruption of Mt. Agung, which is south of the equator, show large optical depths in the Southern Hemisphere, and smaller ones in the Northern Hemisphere. (From Pollack et al. 1976a.)

Knowledge of the sulfate optical depth and its variability comes rather indirectly. Since sulfates seem to constitute about half the world's aerosol mass, they may supply half the global average aerosol optical depth. Indeed, since the optical depth is proportional to the surface area of the aerosols and the mass is proportional to their volume, a fixed mass of small sulfate particles has a much larger optical depth than the same mass of large rock or salt particles. Man's contribution to the sulfate supply is roughly half, or 25% of the total aerosol optical depth. Studies in New York City (Leaderer et al. 1978) suggest that sulfates are responsible for 50% of the light scattering there. Weiss and co-workers (1977) find that sulfates dominate the light scattering in the midwestern and southern United States, and Waggoner and co-workers (1976) reach a similar conclusion about light scattering in Scandinavia. Several pollution episodes of greatly enhanced sulfate levels and reduced visibility over extensive regions of the eastern United States have been mapped by large-scale monitoring networks (Hidy et al. 1978). Annual averages of aerosol optical depths are much higher in central Europe and the eastern United States than in surrounding regions (Flowers et al. 1969; Yamamoto et al. 1968); the difference seems to be due to sulfates most probably

shows much lower tropospheric optical depths at higher latitudes than at lower latitudes, and much lower optical depths over oceans than over continents (Toon and Pollack 1976). There is little evidence for large global-scale changes in the tropospheric optical depth. The optical depth has been monitored at remote mountaintop sites for several decades, and no long-term trends have yet been detected (Roosen et al. 1973; Ellis and Pueschel 1971). However, at many rural sites near cities, industrialized

aerosol concentrations, but in several years we expect to have satellite-borne lasers that will be able to map tropospheric aerosol concentrations and greatly improve our knowledge.

Of the major aerosol constituents—sea salt, soil, and sulfates—large optical depth changes due to sea salt are the least likely, whereas changes due to soil particles are the most prevalent and best documented. Significant regional soil aerosol enhancements result from dust storms near desert produced by humans (Weiss et al. 1977). Trijonis (1979) has empirically linked anthropogenic sulfate emissions with increasing aerosol optical depth in the American Southwest.

Changes in climate

Regional and global changes in climate have yet to be explained satisfactorily. The radiative impact of stratospheric aerosols during the year or two following a volcanic eruption is probably the least controversial agent of climatic change, because the eruption serves as a time marker after which the Earth's climate can be carefully monitored for small changes until the volcanic debris has fallen from the sky. Large volcanic eruptions can provide a unique, well-defined check upon climate models. A major difficulty in testing climate models or in determining the cause of climate changes is that most agents, such as CO_2, tropospheric aerosols, or the sun's luminosity, vary rather slowly and thus their effects can be confused with each other and with natural fluctuations.

Benjamin Franklin first suggested that volcanoes might affect weather when he ascribed the harsh winter of 1784 to a "dry fog" whose origin he thought might have been a large volcanic eruption. Since Franklin's time, several people have studied the weather after an explosive eruption to see if they could discover an effect (Lamb 1970). A statistical correlation was found which suggested that an eruption cools the Earth by several tenths of a degree. Sometimes severe weather changes after a large eruption have caused much human suffering, as in 1816—the "year without a summer"—when frost and cold plagued New England and Western Europe (Hoyt 1958). Some of the suffering of 1816 still remains to haunt moviegoers, since the bad weather that year provided Mary Shelley the opportunity to write *Frankenstein*.

The only violent explosive volcanic eruption since 1912 was the eruption of Mt. Agung, in Indonesia, in 1963 (see Fig. 6), and thus most of our efforts to relate stratospheric aerosols and climate are based on very old data. The data from the Mt. Agung eruption indicate that the dust from that volcano was mainly restricted to

the tropical Southern Hemisphere. A 0.5°C cooling was observed in the tropics at the surface, whereas the stratosphere was observed to warm by several degrees. The empirically observed link between the atmosphere and the aerosols can be tested by calculating the climatic change that should have been caused by the observed aerosols. The calculation is based upon a sophisticated radiative-balance model (Fig. 7) including very simple dynamics (Hansen et al. 1978). Despite its basic simplicity, the

us added confidence that aerosols do affect climate and also helps to reassure us that we may one day actually be able to predict climate changes. Of course, climate models will require much further development and testing before we can use them with assurance.

Several studies based on observations of stratospheric volcanic optical depths (see Fig. 6) over the last century have suggested that some of the observed global mean temperature

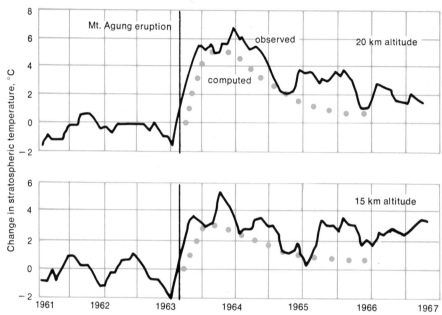

Figure 7. After the Mt. Agung eruption, the stratospheric temperatures, measured at two altitudes, were found to increase due to the aerosols' absorption of solar and infrared radiation. The surface temperature in the tropics decreased because the aerosols reflected some sunlight back to space and absorbed some sunlight that would normally have reached the ground. The time evolution of the temperature partly reflects the changing aerosol optical depths and partly the response time of the troposphere and the ocean. The calculation, which was based on observed aerosol optical depths, is a one-dimensional radiative model and not just an empirical fit. It is encouraging to find that a simple climate model correctly calculates the observed temperature change. (From Hansen et al. 1978.)

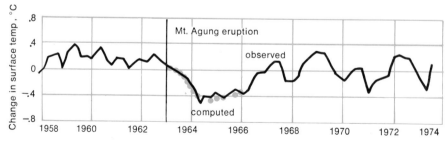

results of the calculation correspond quite well with the observed cooling at the surface (caused by the volcanic aerosols scattering sunlight back to space) and the observed warming in the stratosphere (caused by the aerosols absorbing infrared and solar radiation). This correspondence gives

changes were due to the changing level of volcanic activity (Robock 1978). Figure 8 presents the results of radiative-balance calculations of the Earth's global mean surface temperature in the Northern Hemisphere over the last century. Part of the change of temperature is due to the increasing levels of CO_2 in the atmosphere, and part is due to the decline in volcanic optical depth after 1912 (Pollack et al. 1976a, b). The calculations suggest that volcanic activity from 1880 to 1910 was more significant than CO_2 in causing the temperature difference between 1880 and 1940. Increasing carbon dioxide levels, which have had only a slight effect on the climate so far, will be much more significant in the future (Williams 1978).

Our knowledge of volcanic activity

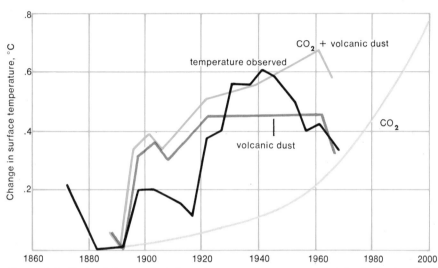

Figure 8. Radiative-balance calculations determine the temperature changes (recorded between 0° and 80° N) over the last century that may have been caused by the observed increase in CO_2, the observed decrease in dust from volcanic activity, and the combined changes from CO_2 plus volcanic dust. Comparison of the calculated and observed temperatures suggests that the decline in volcanic activity made a major contribution to the observed climate warming in the Northern Hemisphere between 1880 and 1940. Estimates show that CO_2 warming may become significant during the next several decades. (Pollack et al. 1976b.)

phase of the last ice age, 20,000 years ago (Gow and Williamson 1971). However, we have no evidence of a significant burst of volcanic activity 100,000 years ago at the initiation of the last ice age, and the periodicity of ice ages strongly suggests that they are caused, at least in part, by variations in the Earth's orbit about the sun (Hays et al. 1976).

Perhaps the most famous theory connecting tropospheric aerosols and climate is R. Bryson's "human volcano" hypothesis (1972) that the

and climate before 1880 becomes increasingly poor the further back we go. The large volcanic optical depths between 1500 and 1900 suggested by Lamb (1970) may have played a causative role in the cool temperatures of that period, which has been called the "little ice age" (Pollack et al. 1976b). However, it is also possible that the climate was influenced by solar-luminosity fluctuations during those centuries (Eddy 1977; Pollack et al. 1979) or by surface albedo changes caused by deforestation and desertification (Sagan et al. 1980). Several scientists have suggested that volcanic eruptions are correlated with ice ages (Bray 1977), and it is true that there was a significant amount of volcanic activity during the coldest

0.3°C cooling which has been observed between 1940 and 1970 (see Fig. 8) is due to aerosols injected into the atmosphere by humans. Figure 3 suggests that if about 25% of the present aerosol optical depth were due to anthropogenic aerosols with a high value of $\tilde{\omega}_0$, they could be the cause of the observed cooling. But we do not even know if the globally averaged optical depth has changed by such a small amount as 25%, although the optical depths in many regions are known to have changed significantly since about 1940 (Yamamoto et al. 1971; Husar et al. 1979; Machta 1972; Trijonis 1979). The human contribution to tropospheric sulfates at present is probably great enough to have increased the optical depth by

25%, though the timing of the change is not known. Additional contributions to the optical depth could have been made by agricultural activities—one of Bryson's favorite ideas. Other possible causes of the temperature change since 1940 include surface albedo changes from alterations in human land use (Sagan et al. 1980) and natural, random climate fluctuations (Robock 1978).

The possible effects of tropospheric aerosols are more easily observed on a regional scale. Bryson (1972) has suggested that the dusty Rajputana Desert of India, site of the ancient Indus Valley civilization, is an example of an anthropogenic aerosol desert. He believes that overgrazing of arid lands by the domesticated animals of the Rajputana inhabitants allows the wind to blow large quantities of dust into the atmosphere; the dust increases the infrared radiation from the lower atmosphere, causing the air to cool and subside, which suppresses rainfall (rain normally falls only in ascending air parcels). Harshvardhan and Cess (1978) have reexamined this problem with a better radiative transfer scheme, and they find that the dust contributes negligibly to the infrared cooling of the air. As yet no one has considered the impact of the dust on the solar radiation, and whether the dust in the atmosphere over the Rajputana Desert is partly responsible for creating the desert remains to be determined by further study.

Another good candidate for empirical testing of the effects of aerosols on climate is Sahara dust over the Atlantic Ocean. The radiative properties of this dust have been characterized by Carlson and Caverly (1977), who suggested that the dust should lead to cooling at the surface and increased solar heating in the atmosphere above the ocean. The significance of these effects to the weather over the ocean has not yet been investigated.

Husar and co-workers (1979) have attempted to correlate for each season of the year observed increases in aerosol levels over the eastern United States with regional climate changes observed since 1948. Of the seasons and regions considered, they found that the aerosol levels increased most dramatically in the Smoky Mountain region during summertime. Simultaneously the same region also experienced the largest changes of any region in several climatological variables, including higher humidity and lower temperature, with noontime temperature decreasing by about 1°C.

One might imagine, because of the obvious importance of aerosols to visibility in urban areas, that the radiative effects of aerosols on the urban climate would be well known. But in addition to the large quantities of aerosols, cities also have large concentrations of gaseous absorbers, they release large quantities of heat into the environment, and their surface properties, such as ground-heat capacity and reflectivity, differ substantially from those of surrounding rural areas. Theoretical studies (Ackerman 1977) have shown that the radiative impact of aerosols on the urban climate is less important than the release of heat or the urban modification of surface properties. Although aerosols do greatly modify the surface radiation energy budget, these modifications seem to be partly offset by changes in other heat transfer processes, such as latent and sensible heat transfer by atmospheric motions.

The most significant radiative impact of the aerosols in urban areas seems to be that they warm the atmosphere and stabilize it against convection. Mixing processes that remove pollutants from urban areas are highly sensitive to the stability of the atmosphere. Urban dwellers quickly learn that temperature inversions, which cause the atmosphere to be stable, lead to high-pollution episodes. Although aerosols tend to promote such inversions, they are primarily caused by large-scale meteorological processes. Aerosols do seem to affect the climate of cities, but because other processes have a stronger influence, it is necessary to make detailed empirical studies that can ascertain the aerosol-climate relations.

The empirical and theoretical evidence for a cause-and-effect relationship between increased levels of stratospheric aerosols due to volcanic explosions and a cooling of the Earth's surface is the most secure of any relationship that has yet been explored by climatologists. However, much further experimental work

needs to be done to determine more precisely the aerosol size distribution and the magnitude and types of climatic changes that occur after large eruptions. The climate shifts caused by large explosive volcanic eruptions can potentially be used to test climate models.

Although the empirical and theoretical evidence for a relationship between increased levels of tropospheric aerosols and climate change is tentative, it suggests that aerosols have affected the climate in many parts of the Earth. Several regions have high tropospheric aerosol levels, and further studies in these regions are needed. Adding nearly transparent materials to the lower atmosphere, such as sulfates and most soil particles, tends to cool the Earth's surface. Adding opaque materials to the atmosphere, such as soot, tends to warm the atmosphere. Since human activity is adding soot, sulfates, and soil to the lower atmosphere in different regions, some areas are probably being warmed and some are being cooled.

References

Ackerman, T. P. 1977. A model of the effect of aerosols on urban climates with particular application to the Los Angeles Basin. *J. Atmos. Sci.* 34:531–47.

Bach, W. 1976. Global air pollution and climatic change. *Rev. Geophys. Space Phys.* 14:492–73.

Bergstrom, R. W. 1973. Extinction and absorption coefficients of the atmospheric aerosol as a function of particle size. *Beit. Phys. Atmos.* 46:223–34.

Bray, J. R. 1977. Pleistocene volcanism and glacial initiation. *Science* 197:251–54.

Bryson, R. A. 1972. Climate modification by air pollution. *The Environmental Future*, ed. N. Polunin. London: MacMillan.

Carlson, T. N., and J. M. Prospero. 1972. The large-scale movement of Saharan air outbreaks over the northern equatorial Atlantic. *J. Appl. Meteor.* 11:283–97.

Carlson, T. N., and T. S. Caverly. 1977. Radiative characteristics of Sahara dust at solar wavelengths. *J. Geophys. Res.* 82:3141–52.

Eddy, J. A. 1977. Climate and the changing sun. *Climatic Change* 1:173–90.

Ellis, H. T., and R. F. Pueschel. 1971. Solar radiation: Absence of air pollution trends at Mauna Loa. *Science* 172:845–46.

Flowers, E. C., R. A. McCormick, and K. R. Kurfis. 1969. Atmospheric turbidity over the United States, 1961–1966. *J. Appl. Meteor.* 8:955–62.

Gow, A. J., and T. Williamson. 1971. Volcanic ash in the Antarctic ice sheet and its possible climatic implications. *Earth Plant. Sci. Lett.* 13:210–18.

Hansen, J. E., and L. D. Travis. 1974. Light scattering in planetary atmospheres. *Space Sci. Rev.* 16:527–610.

Hansen, J. W., W. C. Wang, and A. A. Lacis. 1978. Mount Agung eruption provides test of a global climate perturbation. *Science* 199:1065–68.

———. 1979. Climatic effects of atmospheric aerosols. *Proc. Conf. on Aerosols: Urban and Rural Characteristics, Source and Transport Studies.* New York Academy of Sciences, NY, Jan. 1979.

Harshvardhan and R. D. Cess. 1978. Effect of tropospheric aerosols upon atmospheric infrared cooling rates. *J. Quant. Spect. Rad. Trans.* 19:621–32.

Hays, J. D., K. Imbrie, and N. J. Shackleton. 1976. Variations in the Earth's orbit: Pacemaker of the ice ages. *Science* 194:1121–32.

Herman, B., R. S. Browning, and J. J. De Luisi. 1975. Determination of the effective imaginary term of the complex refractive index of atmospheric dust by remote sensing: The diffuse direct method. *J. Atmos. Sci.* 32:918–25.

Hidy, G. M., P. K. Mueller, and E. Y. Tong. 1978. Spatial and temporal distributions of airborne sulfate in parts of the United States. *Atmos. Environ.* 12:735–52.

Hoyt, J. B. 1958. The cold summer of 1816. *Assoc. Am. Geog.* 48:118–31.

Husar, R. B., D. B. Patterson, J. M. Holloway, W. E. Wilson, and T. G. Ellestad. 1979. Trends of Eastern U.S. haziness since 1948. *Proc. Fourth Symp. on Atmos. Turbulence, Diffusion and Air Pollution*, pp. 249–56. Reno, NV, Jan. 1979.

Junge, C. E. 1972. Our knowledge of the physico-chemistry of aerosols in the undisturbed marine environment. *J. Geophys. Res.* 77: 5183–211.

Lamb, H. H. 1970. Volcanic dust in the atmosphere; with a chronology and assessment of its meteorological significance. *Phil. Trans. Roy. Soc. London* 266:425–533.

Leaderer, B., et al. 1978. Summary of the New York summer aerosol study. *J. Air Poll. Control Assoc.* 28:321–27.

Lindberg, J. 1975. The composition and optical absorption coefficient of atmospheric particular matter. *Quant. Elect.* 7:131–39.

Machta, L. 1972. Mauna Loa and global trends in air quality. *Bull. Amer. Meteor. Soc.* 53: 402–20.

Mitchell, J. M., Jr. 1970. A preliminary assessment of atmospheric pollution as a cause of long-term changes of global temperature. In *The Changing Global Environment*, ed. S. F. Singer, pp. 149–75. Reidel.

Palmer, K. F., and D. Williams. 1975. Optical constants of sulfuric acid: Application to the clouds of Venus? *Appl. Opt.* 14:208–19.

Patterson, E., D. A. Gillette, and B. H. Stockton. 1977. Complex index of refraction between 320 and 700 nm for Sahara aerosols. *J. Geophys. Res.* 82:3153–60.

Pollack, J. B., W. J. Borucki, and O. B. Toon. 1979. Solar spectral variations: A drive for climatic changes? *Nature* 282:600–603.

Pollack, J. B., O. B. Toon, C. Sagan, A. Summers, B. Baldwin, and W. Van Camp. 1976a. Volcanic explosions and climatic change: A theoretical assessment. *J. Geophys. Res.* 81:1071–83.

_____. 1976b. Stratospheric aerosols and climatic change. *Nature* 263:551–55.

Pollack, J. B., O. B. Toon, A. Summers, W. Van Camp, and B. Baldwin. 1976c. Estimates of the climatic impact of aerosols produced by space shuttles, SST's and other high-flying aircraft. *J. Appl. Meteor.* 15:247–58.

Prospero, J. M., and R. T. Nees. 1977. Dust concentration in the atmosphere of the equatorial north Atlantic: Possible relationship to the Sahelian drought. *Science* 196:1196–98.

Ramanathan, V., and J. A. Coakley, Jr. 1978. Climate modeling through radiative-convective models. *Rev. Geophys. Space Phys.* 16:465–89.

Robock, A. 1978. Internally and externally caused climate change. *J. Atmos. Sci.* 35:1111–22.

Roosen, R. G., R. J. Angionne, and C. H. Klemcke. 1973. Worldwide variations in atmospheric transmission observations. *Bull. Am. Meteor. Soc.* 54:307–16.

Rosen, H., A. D. A. Hansen, L. Gundel, and T. Novakov. 1978. Identification of the optically absorbing component in urban aerosols. *Appl. Opt.* 17:3859–61.

Sagan, C., O. B. Toon, and J. B. Pollack. 1980. Human impact on climate: Of significance since the invention of fire? *Science* 206:1323–62.

Toon, O. B., and J. B. Pollack. 1976. A global average model of atmospheric aerosols for radiative transfer calculations. *J. Appl. Meteor.* 15:225–46.

Toon, O. B., J. B. Pollack, and B. N. Khare. 1976. The optical constants of several atmospheric aerosol species: Ammonium sulfate, aluminum oxide and sodium chloride. *J. Geophys. Res.* 81:5733–48.

Trijonis, J. 1979. Visibility in the southwest: An exploration of the historical data base. *Atmos. Environ.* 13:833–43.

Turco, R. P., P. Hamill, O. B. Toon, R. C. Whitten, and C. S. Kiang. 1979. A one-dimensional model describing aerosol formation and evolution in the stratosphere, I: Physical processes and mathematical analogs. *J. Atmos. Sci.* 36:699–717.

Turco, R. P., R. C. Whitten, O. B. Toon, J. B. Pollack, and P. Hamill. 1980. Carbonyl sulfide, stratospheric aerosols, and terrestrial climate. *Nature* 283:283–86.

Twitty, J. T., and J. A. Weinman. 1971. Radiative properties of carbonaceous aerosols. *J. Appl. Meteor.* 10:725–31.

Waggoner, A. P., A. J. Vanderpol, R. J. Charlson, S. Larser, L. Granat and C. Tragardh. 1976. Sulfate-light scattering ratio as an index of the role of sulphur in tropospheric optics. *Nature* 261:120–22.

Weiss, R. E., A. P. Waggoner, R. J. Charlson, and N. C. Ahlquist. 1977. Sulfate aerosol: Its geographic extent in the midwestern and southern United States. *Science* 195:979–81.

Weiss, R. E., A. P. Waggoner, D. L. Thorsell, J. S. Hall, L. A. Riley, and R. J. Charlson. 1978. Studies of the optical, physical, and chemical properties of high-absorbing aerosols. In *Proc. Conf. on Carbonaceous Particles in the Atmosphere*, pp. 257–62. U.C. Berkeley, March 1978. Lawrence Berkeley publication 9037 (available from NTIS).

Williams, J. 1978. *Carbon Dioxide, Climate and Society*. Pergamon Press.

Yamamoto, G., M. Tanaka, and K. Arao. 1968. Hemispherical distribution of turbidity coefficient as estimated from direct solar radiation measurements. *J. Meteor. Soc. Japan* 46:287–300.

Yamamoto, G., M. Tanaka, and K. Arao. 1971. Secular variation of atmospheric turbidity over Japan. *J. Meteorol. Soc. Japan* 49:859–65.

Index